GOLD:

THE SAGA OF THE
EMPIRE MINE
1850–1956

F.W. McQUISTON, JR.

EMPIRE MINE PARK ASSOCIATION
1986

© 1986 Empire Mine Park Association
10787 E. Empire St., Grass Valley, CA 95945

ISBN 0-931892-07-4

Published by Blue Dolphin Publishing, Inc.
Printed in the United States of America by
Blue Dolphin Press, Inc., P.O. Box 1908, Nevada City, CA 95959

10 9 8 7 6 5 4 3 2

TABLE OF CONTENTS

F.W. McQuiston, Jr.

PREFACE

FRANK W. MCQUISTON, JR., a graduate of the College of Mines, University of California, Berkeley, started working in the Empire Mine's assay office in February 1934. He became a metallurgist for the Empire and North Star milling plants in 1937. He began Newmont Mining Corporation's research laboratory, using the available North Star facilities. The Research and Development laboratory was the forerunner of Newmont's worldwide metallurgical research center at Danbury, Connecticut.

While employed at the Empire Mine, McQuiston personally knew George Starr, William Simkins, Arthur B. Foote, J.R.C. Mann, Fred Nobs, and W.D. (Bill) Johnston, a geologist with the U.S. Geological Survey.

With the advent of World War II and the closure of the gold mines, McQuiston became a metallurgist for Newmont's western base-metal mines. After the war he moved to New York where he became chief metallurgist for Newmont's worldwide operation. In 1964 he was elected vice-president of Newmont Mining Corporation. McQuiston served 48 years with Newmont, including 13 years as a consultant. It was during this time that he was instrumental in having the State of California Department of Parks and Recreation acquire the surface rights of the Empire property for the Empire Mine State Historic Park.

This history of the Empire Mine was compiled from many references, including records and maps on file at the mine. Considerable historical data was collected over the years by the author from other sources. Following the acquisition and merger of the

7

Empire and North Star mines, annual reports summarizing the years, activities, and results were issued. It was good fortune to have available a complete set of Annual Reports from 1929 through 1956.

The author wishes to acknowledge the contribution of my old friend Downey Clinch who has made publication of this work possible. Our thanks also to Clarice Truman of Nevada City for her careful reading of the manuscript and glossary recommendations; to Fred Searls III of Nevada City for biographical notes on his father; to Phil Keast and Carl Beyer of Grass Valley for verification of important data; and to Rudy Kopf of Grass Valley for terminology definitions, bibliography correction, and editing assistance. Finally, we thank Ann and Charles Steinfeld, docents at the Empire Mine State Historic Park, for proofreading, editing, restructuring, and augmenting the final text of the manuscript.

F.W. McQuiston, Jr.
San Rafael, California
April 1986

1

GOLD FROM ANTIQUITY

Historical Background

MINING OF PLACER GOLD is as old as the first glimmer of civilization. There is ample evidence that gold had been used for ornaments prior to historic times. Allusions to gold ore are found in the Old Testament. Gold has been sought after and cherished by mankind for more than 6000 years, not just by the prospector, but by kings, governments, conquerors, adventurers, and corporations. All five continents and many of the world's islands have been searched and mined for gold.

Records indicate that some 3 billion ounces of gold have been mined since Columbus discovered America, with more than half mined since 1900. Historians estimate that somewhat less than 1 billion ounces were mined in the 4500 years before the year 1500, making a total of 4 billion ounces mined in the 6000 years since gold was known to be mined in any quantity.

Gold has played an important part throughout the history of mankind—not only due to its monetary value, but for its artistic and unique physical properties. Gold is a yellow metal with a specific gravity of 19.3. Its chemical symbol is Au. It is so malleable that 1 ounce can be pounded into a sheet 100 feet square, or drawn into 50 miles of wire. Gold is found naturally in several forms, the most common are the alloys electrum and amalgam and the telluride compounds. However, gold usually occurs as a relatively pure metal.

Egypt was the main source of gold in the ancient world, being the richest gold-producing country for many years from both placers and shallow vein deposits. Mining of gold from hard rock was an ancient art also practiced by the Egyptians, the first recorded miners. The rock was broken by heating with a wood fire built against the underground face, followed by quenching with cold water which caused the rock to crack. The cracked ore was pried out with a moil and hammer and carried to the surface in buckets or in skins by gangs of slaves. It was then broken down to pea size and ground to fine sand grains in granite mortars. The ground material was washed over inclined tables on which the particles of gold were caught on the rough surfaces. The Egyptians probably mined lode gold for thousands of years, and the remains of their gold mines can still be seen in the desert between the Red Sea and the Nile River. In one mine it is estimated that 1 million tons of rock were excavated. At another mine 4 miles of tunnels are still intact. Excavations of early Egyptian tombs disclosed that the art of goldsmithing was highly developed.

During the first several centuries A.D., gold was mined in Europe, Africa, Russia, Mexico, and South and Central America. Gold has been the symbol of wealth in all civilizations that have arisen in the past 10,000 years. The Assyrians, the Egyptian dynasties, the ancient empires of India and China, Greece and Rome, the Incas, and the Aztecs have all worshipped at the feet of the Golden Calf. Throughout history men have toiled, fought, and died for this beautiful and enduring metal.

Gold was important in the days of Marco Polo's travels, and it was a major trade commodity during the Roman Empire. The gold of South and Central America was a major factor in the exploitation of the New World by the Spanish Conquistadors in the 16th and 17th centuries.

History of Gold Metallurgy

OUR KNOWLEDGE OF MODERN-DAY gold and silver metallurgy is based on the foundation of Georgius Agricola's De Re Metallica (Herbert Hoover Translation, 1950 Edition, p. 354) published in 1556. Agricola was closely associated with one of God's gifts of great mines, Joachimsthal, Czechoslovakia. The knowledge set forth in Metallica would have been extremely helpful to our early day miners had the translation been available.

Quicksilver (mercury), produced from cinnabar, appears to have been used for amalgamation of gold in Roman times. Gold metallurgy had its inception when millstones were used to grind gold-bearing quartz with mercury, thus forming gold amalgam. This was followed by the mechanized stamp mill in 1519 in Joachimsthal. The drawings in De Re Metallica of the stamps, riffles, strakes, and sluice boxes were quite similar to those developed independently and used by early American gold miners.

Georgius Agricola published the following tabulation:

Gold washed from alluvial: Prior to recorded civilization.

Gold reduced from ores by concentration: Prior to 2500 B.C.

Gold refined by cupellation: Prior to 500 B.C.

Gold recovered by amalgamation: Prior to Christian Era.

The oldest implement for the recovery of gold used in ancient times consisted of a trough of boards several feet long, with holes in the bottom to let sand and small gravel fall through into a box below. The material collected in the box was then placed in bowls and the gold was separated by panning.

A second implement consisted of a trough with riffles across it. After the riffles had caught enough material to fill them, they were removed and the material collected in bowls. Variations in sluices used grooves in the planks to catch the heavier particles and cloth or the skins of animals (hence the term "golden fleece").

Although the recovery of gold by amalgamation dates from Roman times, silver amalgamation was not used prior to the 16th century. Parting of gold and silver was known in the Middle Ages between 1100 and 1500 A.D.

The first written description of gold occurring in veins rather than in placer deposits appeared in the second century B.C. Crushing by hand continued until the stamp mill was invented in the 15th or early 16th century. Amalgamation followed by concentration and smelting of the concentrate with litharge or lead ores and cupellation of the lead for gold and silver recovery was used until hydrometallurgical processes were invented.

"Cupellation" is a step in the process of separating metals, practiced in ancient times. It is still used in silver refineries and is an important step in modern assaying. A cupel is a small cup or thimble made of bone ash. When a mixture of gold and lead is heated to melting in a cupel, the lead oxide formed by the oxidizing atmosphere is absorbed in the walls of the cupel, leaving a button of gold and silver in the bottom.

Gold Discovered in the United States

GOLD WAS FIRST DISCOVERED in the United States in North Carolina in 1801. Prior to the California discovery in 1848, approximately 7500 ounces of gold had been mined in several of the southeastern states. Earlier gold production in Southern California by Mexican prospectors was inconsequential.

California produced 13 million ounces from 1851 to 1855 and reached peak production of 3 million ounces in 1853. The United States became the leading producer in the world, a position it maintained for 50 years due to later discoveries, notably: the Comstock Lode, Nevada, in 1859; the Colorado fields in 1860; Homestake Mine, South Dakota, in 1877; and Cripple Creek, Colorado, in 1891.

The discovery of gold in California was the start of the modern Golden Age in mining with far-reaching consequences. The search for gold fields was vigorous with the finding of gold in Australia as a direct result.

The World's Greatest Gold Discovery

THE MOST MONUMENTAL DISCOVERY of all times was made in February 1888 of gold-bearing conglomerate rock on a farm near Johannesberg, Transvaal, South Africa, known the world over as the Witwatersrand or the "Rand."

Everything concerning gold discovery, mining, and production paled before the extraordinarily extensive and productive deposits which have been subsequently developed on the Rand and the Orange Free State. With a total production of over 1.2 billion ounces since its discovery almost 100 years ago, the Rand has already produced about one-third of the world's total production.

Some of the mines have been developed to over 2.5 miles in depth. The translation into production of the deep-going Rand deposits deserves to rank as one of the major achievements in mining history and as one of the greatest demonstrations of engineering skill in the mining profession.

Early Treatment of Gold Ore in the United States

IN 1829 THE FIRST STAMP MILL in the United States was in operation in the state of Georgia. At that time amalgamation was used for the recovery of gold from the crushed ore. In 1850, in California, the stamp mill again was used; it was then supposed that crushing and amalgamation was the only process suitable for treating gold ores, and gold remaining in the tailings from this process was generally regarded as not recoverable. Later it was found that a concentrate could be made from the mill tailings and shipped profitably to a smelter.

In 1848 chlorination was first used to process arsenical ores in Silesia, Czechoslovakia. Chlorine, in the presence of moisture, converts gold into the tri-chloride, which is soluble in water; it can then be separated from the ore by filtration and the gold precipitated by various methods. Sulfide ores or concentrates were roasted before chlorination.

At first, chlorine treatment was done in small barrels or vats with slats at the bottom. A layer of perforated boards was covered with a layer of gravel and sand that acted as a filter, and the moist crushed ore was loosely charged on top of the sand. Gaseous chlorine was fed into the bottom of the vat with a contact time of from 12 to 36 hours. This process was particularly well adapted to the treatment of roasted telluride ores or auriferous sulfides. It was first used in Grass Valley in 1858.

Discovery of Cyanidation

THE MOST IMPORTANT DISCOVERY concerning the treatment of gold ores was made by L. Elsner, a German chemist, in 1846, when he demonstrated the solubility of gold in dilute solutions of potassium cyanide, but he failed to recognize its significance. J.S. MacArthur, a metallurgical chemist in Glasgow, Scotland, realized that Elsner's discovery might be of value, and in conjunction with R.W. Forrest and W. Forrest, doctors of medicine, developed a process using cyanide for recovery of gold. In 1887 they registered their first patent in Great Britain. Their extraction and precipitation processes were patented in the United States in 1889.

Within a few years cyanidation plants were in operation in gold mining districts throughout the world and led to several very important inventions useful not only in cyanidation plants but in many other ore processing operations. Since the degree of recovery varies depending on the type of ore, there is no single process adaptable to all ores. Much depends upon the gangue, other metallic minerals present, and the distribution of the gold in the matrix. If coarse free gold is present, it may be recovered by gravity methods or by amalgamation. In some cases cyanidation is conducted separately on the sand and slime fractions; in other cases it is preferable to grind the ore so finely that cyanidation can take place during agitation.

2

HISTORICAL REVIEW OF CALIFORNIA MINING

First Discovery of Gold (Placer Gold)

THE FIRST DISCOVERY OF GOLD in the Sierra Nevada is credited to J.W. Marshall, who in January 1848 found nuggets in the millrace while constructing a saw mill at Coloma, El Dorado County. Placer gold was first found in Wolf Creek, Grass Valley, in 1848, shortly after Marshall's discovery, and the discovery of rich placers in 1850 in Deer Creek, near the site of Nevada City, brought the first gold rush to that area.

Very soon, however, the older Tertiary Hill gravels were discovered; these deposits were far richer and more abundant in Nevada City than in Grass Valley. The term "Tertiary Hill gravels" identifies the rich, gold-bearing river beds which were thrust upwards in the course of the mountain forming activity of the Tertiary Period, a period in the Earth's geologic time scale running from approximately 70 million to 25 million years ago. It was the period that saw the formation of the Cascade, the Sierra, and the Rocky Mountain ranges.

Hydraulic mining was successfuly introduced in the huge Tertiary gravel banks of the ancient river beds, but was abandoned in 1884 when the courts ruled that this method of mining could be used only if a positive means could be established to prevent the resultant debris from entering the rivers and streams.

Discovery of Lode Gold and the Empire Mine

THE FIRST FIND OF GOLD-BEARING QUARTZ was the discovery of the Gold Hill ledge in June 1850 by George McKnight. The quartz in

14

this outcropping was literally filled with gold, demonstrating clearly the source of the rich nearby stream deposits.

The richness of the find caused a flurry of intense prospecting which, in October 1850, led to the discovery of gold-bearing ledges on Massachusetts Hill, Ophir Hill, Rich Hill, and other strikes. It was the Ophir Hill claim, located by George Roberts, which subsequently became the Empire Mine. So many of these early claims contained gold ore so rich and plentiful that, for a time, the mine owners feared gold would lose its value.

Because of the richness of the ore and the fact that a satisfactory method for recovering the gold still needed to be developed, the early claims were only 30 by 40 feet.

These discoveries marked the advent of new mining techniques for the California miners, at least new and strange to the majority of the men working the placer deposits of the northern rivers. Hard rock mining presented totally different problems. No individual, or even several men, could make a go of it with muscle alone. Pick, shovel, and gold pan—the usual equipment—were inadequate. Mining quartz veins required heavy machinery, explosives, the building of superstructures, an understanding of pumps, and methods of shoring up underground excavations. The capital this required demanded pooling of resources by groups, not only of miners but of outside investors.

It was a learning process for the miners with respect to new mining methods and equipment, and for the investors to reach awareness of the profit possibilities in properly equipped, well-managed operations. The risks were great, and there were trial, error, and disappointment which account, to some degree, for the frequency with which many early claims changed hands.

In spite of all difficulties, the Sierras held the promise of such abundant wealth that it was not long before lode mines flourished up and down the foothills. Over the years the gold-quartz mines of the Grass Valley-Nevada City district became the most productive in the state of California, eventually ranking as the third most productive in the United States, after Lead, South Dakota (the Homestake), and Cripple Creek, Colorado.

During the Civil War, gold from California and Nevada aided materially in the preservation of the Union with a supply of new wealth.

Photo Credit: Hollis DeVines

Giant 30-foot-diameter Pelton Wheel at North Star Mine.

Mining Equipment Development

FOR A LONG TIME AFTER quartz mining started, steam engines using local forest wood for fuel were the principal source of power for the mines and mills. The development of high-level ditches and reservoirs for hydraulic gravel mining and the increasing cost and scarcity of wood led to the use of water power under high pressure to rotate tangential or impulse wheels. By the end of the 1880s, nearly all of the principal quartz mines used water wheels of the impulse type to supply their power needs. The Pelton Water Wheel was the most successful. This wheel was highly efficient and made possible the introduction of electric power in the mining regions of the West before public utilities had entered the field. The first recorded use of electric power for the operation of mining or quartz milling machinery in California was in February 1890.

Hand drills and black powder were used entirely until 1868, when the first air drills and dynamite were introduced. Although there was widespread use of air drills for stoping, they did not come into general mining use for another 30 years or so. In 1906 acetylene lamps replaced the candle, which had been used underground for over 50 years.

Pneumatic drilling in a stope
—North Star Mine.

Photo Credit: Arthur B. Foote photo collection

Loading ore car at the foot of a stope.

Even though the Empire mine used equipment compatible with the latest methods for mining deep narrow veins, many mules were used underground throughout the life of the mine. The mules lived in underground stables and were brought to the surface only when too old to work.

On their way through a drift to the main shaft. Man and mule work as a team.

Typical Cornish miner.

Although successful processing of gold ore, in any quantity, could never have been accomplished without the crushing capabilities provided by the stamp mill, deep hard-rock mining would have been impossible in the early days without the Cornish pump. Developed in Cornwall, England, the Cornish pump was used to clear water from tin, copper, and coal mines.

The first Cornish pump in the West was built and installed at the Gold Hill Mine, Grass Valley, in 1855. Similar pumps were used throughout California and on the Comstock in Nevada. In fact, where water had to be lifted from deep-shaft mines, they were indispensable.

Stamp mill batteries—North Star Mine.

Removing gold amalgam from mortar of stamp mill and amalgamation table—Empire Mine.

The pump rod of the Cornish pump at the Empire was about a half-mile long, had an 8-foot stroke, and operated 8 to 9 strokes a minute. As much as 18,000 gallons of water per hour were delivered to drain tunnels with this pump. The pump at the North Star was of similar length and operated 24 hours a day for 40 years.

For their day these pumps were marvels of innovative efficiency and reliability, but with the development of modern, submersible electric motors and improved pumps, these mechanical giants were phased out and by 1920 were extinct.

Photo Credit: Hollis DeVines

Cornish pump at North Star Museum.

3

GEOLOGY

Grass Valley Area Veins

LINDGREN (1896) MAPPED the surface geology and described the ore deposits, and Johnston updated the mine data in 1940. The following geologic description is extracted from Johnston's paper.

The major geologic feature of the district is a body of early Cretaceous granodiorite 5 miles long from north to south and from a half-mile to 2 miles wide which has intruded older rocks and is itself cut by various dike rocks.

The oldest rocks of the district belong to the carboniferous Calaveras formation, formerly clastic sediments but converted to schistose and slaty rocks during a late Paleozoic orogeny. Igneous rocks other than the granodiorite include aplite, diorite, gabbro, and diabase. Some serpentine and amphibolite schists are present in the northeast part of the area.

Gold-bearing quartz veins occur in or are associated with all of the rock units previously mentioned. After the gold-bearing veins were formed, the area was uplifted, deeply eroded, and buried under Tertiary auriferous gravels and widespread rhyolite-andesite tuffs. Quarternary alluvium is largely confined to present drainage basins.

Most of the veins strike north, parallel to the long axis of the granodiorite body, and have shallow dips averaging about 35°. The most productive veins dip into the granodiorite from both east and west sides and tend to converge with depth. The veins fill minor thrust faults within fracture zones of variable width and degree of shattering. Reverse displacement rarely exceeds 20 feet.

Another important structural feature in the district is a group of vertical or steeply dipping fractures called "crossings" that strike northeast about normal to the long axis of the granodiorite body. Few crossings contain quartz. They are significant in an economic sense because they commonly bound ore shoots. By breaking the veins into segments, each of which has opened or closed more or less independently of its adjacent segments, the crossings have permitted mineralization in certain open segments, whereas other segments which remained closed are barren.

Quartz is the principal vein mineral, and usually several generations of quartz are recognizable. Gold occurs both in quartz and in cracks in broken sulfides, principally pyrite. Although the average value of ores from the district is near 0.6 ounce per ton, the distribution of gold in the ore shoots is extremely erratic, and assays of samples from adjacent veins commonly differ widely. Some ore shoots have a pitch length of several thousand feet, but most are much shorter. There has been no recognizable change in the character of the veins or in the occurrence or distribution of gold through a vertical range of more than 4000 feet, and there is no reason to suppose that mining has reached the bottom of the lowest ore body.

Description of Veins

IN THE GRASS VALLEY DISTRICT, over 100 gold-bearing quartz veins are distributed relatively uniformly over an area of about 10 square miles. The deepest mine workings, nearly a mile vertically below the surface, indicate neither change in the strength and character of the veins, nor change in the character of the mineralization with increasing depth.

The vein minerals were deposited by dilute aqueous solutions, wholly of magmatic origin, in open fissures. Ore shoots are independent of the chemical composition of the country rock. Veins occur in all pre-Tertiary rock units and commonly cross rock contacts without change. The physical character of the wall rock is important only insofar as it controls fracturing and so affords channels for mineralization.

Within the trough defined by the intersecting North Star and Empire veins, there has been enough mining and exploration to discount the potential for very large tonnages of undiscovered ore. In the footwall of this trough, however, exploration has not been extensive, but there is no reason to suspect that the shallow dipping, conjugate veins do not continue to be repeated.

Veins mined have the following average characteristics based on mining experience:

True thickness	16 inches
Strike length	3000 feet
Dip length	2000 feet
Dip angle	30°
Ore content of the vein	30% of total volume
Grade of the ore, in place	1 ounce gold per ton
Vertical interval between veins	500 feet

4

MINING OF THE EMPIRE VEINS

THE EMPIRE VEIN OUTCROPS in the diabase-porphyry and continues to a depth of 1700 feet, where it enters granodiorite which forms both walls of the vein. The average dip of the Empire ledge to the 3000-foot level is 30°, but below 3400 feet the vein steepens, and from the 3800-foot level to the 4200-foot level, the average dip is 55°. The walls of the fissure are 3 to 4 feet apart, but the auriferous quartz averages 18 inches to 2 feet. In the lower levels, however, a drift has been driven a distance of 200 feet on an ore-body with an average width of 8 feet of heavily mineralized quartz. The ore consists of quartz carrying fine and coarse gold with a content of 2% to 3% sulphides, which are essentially auriferous pyrite with small amounts of finely disseminated galena. The sulphides vary in value from 3 to 7 ounces of gold per ton.

The Empire vein was remarkably persistent. It has been followed down dip for over 7000 feet and along the strike for over 5000 feet. Its strike averages about due north and it dips about 35° to the west.

The veins in the Empire property were developed primarily by inclined shafts driven on the dip of the vein. Vertical shafts were sunk in several places. Levels were established at 300- to 400-foot intervals on the incline. At these intervals, drifts 5 by 7 feet in cross-section were driven along the strike of the veins. Raises were driven between the levels in the ore shoots. Open stopes were developed from the raises. The vein, averaging 16 inches in thickness, was blasted separately from the waste to minimize dilution. The height of the stopes averaged 42 inches. Waste was for the most part stored underground in previously mined areas. Timber stulls were used for temporary support. Scrapers were used to move ore and waste from the stope to chutes from which ore cars could be loaded and trammed to shaft loading pockets. Stope productivity was of the order of 3 to 4 tons per man-shift.

Bin on 2500-foot level, Pennsylvania (Empire) Mine at top of winze, showing skip in dump and car under chute.

1400-foot level, Pennsylvania (Empire) Mine, showing method of timbering to protect level from soft hanging wall and vein material.

Double loading platforms, with men shoveling from top platform into car—No. 7 stope, 3600-foot level, Empire Mine.

Go-devil station with empty truck that just came up. Intermediate level 37—No. 1 stope, Empire Mine.

5

EARLY HISTORY
OF THE EMPIRE MINE

Owners and Stamp Mill Problems

THE FIRST 20 YEARS of the Empire's existence saw a number of ownership changes. George D. Roberts held the original Ophir claim for less than a year, selling it in 1851 to Woodbury, Parks and Company who consolidated it with other nearby claims under the "Ophir" name. Ore crushing problems and poor management resulted in sale of the property at auction in 1852. One-half was purchased by John P. Rush and one-half by the Empire Quartz Hill Company.

Early stamp mills were extremely primitive and impractical. The first mill built in November 1850 used stamps made from tree trunks shod with pieces of iron. Other mills using stamps with square wooden stems and square iron shoes failed to perform effectively because there was no way to rotate the stamps, which resulted in very rapid and uneven wear of the shoes. The general difficulty in crushing caused a slump in mining which persisted through 1853. However, operations picked up with development of mills equipped with cylindrical stamp stems of iron fitted with tappets which permitted rotation of each stamp. There followed a rapid evolution of stamp mill design, and by 1857 stamp mills had attained an efficiency which established quartz mining on a broad scale.

All stamp mills were developed from the original concept of Count von Maltiz in Saxony in 1505 that broke and pulverized the various types of quartzitic ores mined. To facilitate the flow of ore and to eliminate dust, water was added to the mortar box. A wet pulp was discharged by the stamps through wire screens onto mercury-coated plates.

Amalgamation of the gold in the pulp was accomplished by passing the pulp over long plates of copper which were first silver plated then coated with a thin layer of liquid mercury.

In 1854 the Empire Quartz Hill Company bought out the interest held by John Rush and the mine was incorporated as the Empire Mining Company. From then on, operations proceeded on a more systematic basis.

The Grass Valley Process

WALDEMAR LINDGREN in his Professional Paper, 1896, p. 23, quoted from W. P. Blake, the process used at the Empire Mill in 1853 as follows:

> The ore after being roasted in heaps, is crushed in a 16-stamp mill; the pulp passes over blankets, where much gold and pyrites is caught; these blankets are wrung out in water at intervals, and the mixture of gold and pyrites is subjected to amalgamation in pans. From the blankets the pulp passes through a revolving cylinder holding mercury where a part of the fine gold is amalgamated; finally the pulp is subjected to an amalgamation in a Blaisdell pan holding mercury and iron balls; three-fourths of the gold is caught on the blankets.

This milling method, known as the Grass Valley Process, was in use for many years.

Production Data, Ownership Change, Frue Vanner Developed

BETWEEN MAY 1854 AND DECEMBER 1863, the Empire Mine produced 28,100 tons of gold-quartz ore which was processed by the stamp mill. The yield of only the free gold was more than 51,000 ounces, for an average of just under 2 ounces per ton, and worth over 1 million dollars at a price of $20 per ton. The production from the discovery in 1850 to May 1854 was estimated to be about 15,000 ounces, worth about $300,000.

In 1864 the Empire was sold to Captain S.W. Lee and A.H. Houston. The new owners made a number of improvements to surface structures and equipment including a new 30-stamp mill, which was considered to be one of the finest mills in the state.

Captain Lee, the principal owner and manager, was a seafaring man who clung to many of his marine ideas. The interior of the mill was ceiled and painted and the exterior so designed that it became known as "The Steamboat Mill."

The sulfide concentrates which made up 2 to 3% of the Empire ore varied in gold content from 3 to 7 ounces, with an average value of $80 per ton, and when recovered were an important source of income. From the first recovery methods using blankets, sluices with riffles, Cornish buddles, and other devices, including new types of mechanical concentrators, came the development of the Frue vanner in 1867. It was an endless belt of rubber with lateral-longitudinal oscillation and proved so effective that the Frue vanner became standard equipment where stamp mills were used. The recovery from the Empire vanners was 14-15% of the total recovered free gold value in the ore.

Chlorination of Sulfides

THE CHLORINATION PROCESS recovered gold from sulfide concentrates. It was known in Europe as the Plattner process and was successfully introduced into the district at the Pioneer Reduction Works, southwest of Nevada City, on December 8, 1858, as a custom plant and by 1890 it had treated 16,000 tons. The average value of the concentrate was about $80 or 4 ounces per ton from which the plant extracted 94%. The reduction works paid approximately 92% of the sample assay value, with a treatment charge of $18 per ton. No allowance for silver was made unless its value exceeded $10 per ton; when this occurred, 60% was returned to the shipper.

The sulfide concentrate was first subjected to an oxidizing roast with salt added in long reverberatory furnaces 16 feet wide by 60 feet long. The reverbs were fueled with pine and cedar wood. After roasting, the sulfide concentrate was trammed to the nine chlorine gassing tubs, each of which was 5.5 feet in diameter by 4.75 feet deep, with an average capacity of 3500 pounds of concentrate.

The works contained two chlorine generators into which manganese, salt, and sulfuric acid were charged. The concentrate in the chlorination tubs was exposed to the chlorine action for 36 to 48 hours. The precipitating vats and leaching vats were coated with a mixture of asphalt and coal tar to prevent corrosion. Gold was precipitated from the leach solution by ferrous sulfate. After the gold was leached out, the residue if sufficiently high in silver was transferred to silver leaching tubs where it was leached with calcium hyposulfite and subsequently precipitated with calcium polysulfide as silver sulfide.

6

A New Owner and
A New Approach to Mining

By 1867 MOST OF THE MAJOR MINES of the district had been located, and in Grass Valley 30 quartz mills were dropping 284 stamps, crushing 72,000 tons per year, with an average yield of $30 to $35 per ton. There were 1600 men working in the industry.

This, however, was the year that saw the beginning of a slump in mining in the Grass Valley-Nevada City area. It deepened into a depression that was to last some years and was of such a magnitude that by 1880 the only mines operating in Grass Valley were the Empire, the Idaho Maryland, and the New York Hill.

The principal causes were widespread bad management, failure to maintain ongoing exploration programs to locate new ore reserves, and a general deterioration in ore quality. Many miners left the area, lured by the rich strikes on the Comstock Lode in Virginia City.

Discouraged mine owners began to look for opportunities to liquidate their investments. In 1867 a half-interest in the Empire was sold to a group of San Francisco investors, including Lake, Cronise, and Horner. It was also the period when William Bowers Bourn, a San Franscisco capitalist who had been buying Empire stock for several years, began to increase his holdings. By 1869 he acquired control of the company.

This was a decisive turning point in the history of the Empire Mine, indicative of a new trend toward ownership by individuals with sufficient financial backing to introduce the best mining equipment and the best available engineers to manage operations.

Quartz mining throughout California began to attain a new stature. It marked a transition from the conduct of mining as an exciting adventure to mining as a serious business. Once the wealth

31

of financial centers across the country began to buy in, hard rock mining became a different game.

The history of the Empire Mine for the next 86 years demonstrated this new stability and bears the mark of a number of highly eminent men in both mining technology and the world of finance.

7

THE EMPIRE UNDER
WILLIAM BOWERS BOURN, SR.

WILLIAM BOWERS BOURN, SR., was born in Somerset, Massachusetts, on June 21, 1813, a descendant of early immigrants who arrived from England in 1630 with the John Winthrop party, founders of the Massachusetts Bay Colony.

As a young man W.B. Bourn entered the banking business, but moved from New York to San Francisco in 1850 to assist his father-in-law, Captain George Chase, in the maritime shipping business. His wife, Sarah Chase Bourn, joined him in 1854, and they made their home in San Francisco. They had six children, two boys and four girls.

After the birth of the last child, Bourn gave his wife a working vineyard and country home in the Napa Valley, called "Madroño," which proved to be a most popular home with all the family for many years.

Shortly after arriving in San Francisco, Bourn reentered the banking business and became active in commission brokerage and trading on the local exchanges. He also began to invest in mining stocks, and in 1869 acquired a controlling interest in the Empire Mine in Grass Valley.

Bourn took a keen interest in operations at the Empire, making the arduous trip to Grass Valley frequently. Just about the time Bourn assumed control of the mine, it experienced its first formal labor problem. Miners went on strike protesting replacement of the black powder they had used for years by the newly developed explosive, dynamite. To ease the situation, the use of black powder was resumed for about a year, but rapidly declined as local experience demonstrated the superiority of dynamite.

In the spring of 1870 Bourn brought in a new mine superintendent, Mr. Nesmith. In September a serious fire totally destroyed the 30-stamp mill erected 5 years previously by Captain Lee. Within a year a new 20-stamp mill of an advanced design was operating, which exceeded the crushing capacity of the old 30-stamp mill. With more milling capacity and with the gradual return of experienced miners from the Comstock Lode in Nevada, the Empire became very prosperous and provided a substantial part of the Bourn family income.

In 1874 Bourn died and the Bourn estate assumed direction of the mine, with Nesmith remaining as superintendent. Operations continued until 1878, when all pay ore of the Ophir vein was extracted, pumping was suspended, and the main incline, which had reached a depth of 1200 feet on the dip of the ledge, was allowed to fill with water. Work was then commenced on a new shaft on a parallel vein, known as the Rich Hill ledge (also discovered in 1850), which had been worked through the Ophir shaft.

A dynamic businessman, intolerant of mediocrity in any form, William Bowers Bourn left an imprint on the Empire which would, subsequently, be broadened and deepened by his son, William Bowers Bourn, Jr.

Empire Production 1865-1878

THE PRODUCTION OF BULLION from the Empire Mine from 1865 to October 1878 was over $1,900,000 or 95,000 ounces of gold. Total production from its founding in 1850 through 1868 was approximately $3.2 million from 162,000 ounces of gold. This amount represents the free gold only and does not include gold recovered from the sulfides.

8

The Empire Under William Bowers Bourn, Jr.

WILLIAM BOWERS BOURN, JR., (1857-1936) was just 17 when his father died. After completing his secondary schooling, he left the next year (1875) for England and Cambridge University. Three years later, at age 21, he left Cambridge and returned to California to help his mother manage the family's numerous business interests.

Critical Years for the Empire

BOURN ARRIVED HOME at a critical period in the Empire's history. His mother and her business advisors were preparing to abandon the mine because it was the opinion of several experts that it had reached a depth (1200 feet) at which mining could no longer be conducted profitably.

Young as he was, Bourn was very much his own man. Refusing to be swayed by those who felt the Empire was a "dead" mine, he checked the workings personally and decided that it still had a good future. Exercising business acumen reminiscent of his father, he reorganized the company, forming the Original Empire Mill and Mining Company which took over all assets of the Empire Mining Company.

Bourn entered the scene when mining throughout the area was in a severe depression, and his first few years were difficult and financially draining. However, pushing ahead with a systematic exploration program, his perseverance was rewarded with the opening of large new bodies of ore which, by 1883, brought a new era of prosperity to the Empire.

Cousin George Starr

NO CHRONICLE OF EITHER THE EMPIRE MINE or the Bourn family would be complete without consideration of George W. Starr, a younger cousin of W.B. Bourn, Jr.

After the death of his mother, George Starr was brought up at the family country home, Madroño, under the guidance of his aunt, Sarah Bourn, who became his legal guardian.

In 1881, when Starr was 19, his cousin brought him to Grass Valley to work in the Empire Mine, and Starr's progress from there on was phenomenal. Possessing an instinct for mining, an engaging personality, and an overwhelming desire to excel at every job, he became exceedingly well-liked by all the miners. Throughout all the later years of the Empire under Bourn family ownership, George Starr proved to be the mine's greatest asset, deserving full credit for its years of successful operation and its contributions to the economy and welfare of the community.

Years of Change and Growth—Empire Sold

THESE WERE YEARS OF EXPANSION and improvement at the Empire and in the mining community generally. Bourn took an increasingly direct role in operations, particularly in long-range planning and began to improve the surface works. With his unique interest in the harnessing of natural resources, in application of new mining techniques, and in new equipment development, the mine moved forward.

The Nevada County Narrow Gauge Railroad, completed in 1876, was used extensively by all the mines, significantly lowering freight costs on equipment and supplies.

In 1878 the Pelton Water Wheel was developed by Lester Pelton of Camptonville, a few miles north of Nevada City. Its radical new design proved it to be a highly efficient source of power. Bourn realized the wood fuel for steam engines was being rapidly exhausted, and laid plans for a changeover to water power.

Building reservoirs, constructing ditches and flumes, laying long pipe lines, and installing the Pelton wheels, each designed for a specific application, were large undertakings, but by 1886 the Empire Mine was operating on water power. The water discharged from the Empire was used twice again before finding its way into Wolf Creek.

*Small
Pelton Wheel
at Empire Mine.*

Photo Credit: Hollis DeVines

A 22-inch line carried it first to the North Star Mine and from there to the Allison Ranch Mine.

By 1880 the stamp mill had been much improved, so that the ore was crushed by rock breakers ahead of the stamps and fed automatically to 850-pound stamps. The same year (1886) the changeover to water power was completed, the number of stamps was increased from 20 to 40. Shortly after this major expansion and modernization program was completed, George Starr, at age 24, became mine superintendent.

In 1884 Bourn purchased the North Star Mine which had been closed for some years, and with the help of John Hays Hammond, a brilliant young mining consultant, it became a major producer. In 1887 Bourn sold the North Star to James D. Hague who represented wealthy eastern investors. A year later Bourn sold a controlling interest in the Empire to Hague, with whom he had developed a strong friendship. George Starr was retained by Hague as manager of the Empire.

New Business Horizons

AT THAT TIME BOURN'S MANY INTERESTS outside of the Empire had multiplied to a point where he decided to retire from mining. Among other undertakings, he had opened a bank in Grass Valley, was a director of the San Franscisco Gas Company, and was starting construction of a huge wine cooperative in St. Helena, about 5 miles from the family home. Called "Greystone," it was built by local stonemasons from native blue-grey volcanic rock and was the largest stone winery in the world. Unfortunately, a vineyard epidemic in 1894 reduced the need for such a large storage capacity and Greystone was sold. Since 1950 it has been owned by the Christian Brothers, a Catholic Order, who are operating it as part of their extensive wine industry.

At the Empire, under Hague's ownership, George Starr continued to improve mining operations, introducing compressed-air drills in 1890 which greatly increased the rate at which drifts, shafts, and development work could be extended. That year, however, they had an extremely severe winter resulting in mine flooding and operating delays. By 1892 the main shaft had reached the 2100-foot level on the incline. The following year the mine encountered a barren zone with a marked decline in profitability.

Starr Goes to Africa—
Bourn Reacquires the Empire

THE HIGH STANDARDS that George Starr continued to demand of himself, his rapid acquisition of mining know-how from the old "Cousin Jacks," and his native intelligence all combined to attract the attention of many notable professional engineers. John Hays Hammond, one of the great mining engineers of the day and consultant during the early years of Bourn, Jr.'s, ownership, was particularly impressed. In 1893 he lured Starr away from the Empire to join him in the goldfields of South Africa, and it was there, in the mines outside Johannesburg, that George Starr matured into a top-flight mine manager.

When Starr left the Empire in 1893 for Africa, Robert Walker took over as superintendent. The barren zone persisted, and with the drop in ore quality the mine began to deteriorate, operating at a loss for several years. In 1896 Bourn reacquired control of the Empire and began, once again, to demonstrate his faith in the mine, as he had 18 years previously. The Empire Mines and Investment Company was formed to take over the assets of the company he had formed in 1878.

The following year he commissioned Willis Polk, a San Francisco architect, to design and build an English manor-style residence known as the "Cottage" on a knoll overlooking the mine. Polk had previously built a palatial town house for Bourn on Webster Street in San Francisco. The Cottage was built largely of waste rock from the mine and trimmed with brick; it had the style of a mansion, but was intended as a summer residence only. However, with a greenhouse, beautiful gardens, fountains, and a reflecting pool, it was a residential showplace in the community.

Following completion of the Cottage in 1898, a clubhouse was constructed nearby with tennis courts, bowling alley, and squash courts—all completed in 1905.

George Starr Returns to the Empire

AFTER FIVE YEARS on the Rand in South Africa, where he had earned as much in a month as he had formerly earned in a year at the Empire, George Starr returned to California in 1898. Cousin Bourn was very anxious to have him resume his managerial role at the Empire and help restore it to its previous prosperity.

It was 10 years since the mine's last major improvement program, and Starr found both buildings and equipment in a deplorable state of decay, the mine having outgrown the plant. He agreed to return only if he was provided with a fund of $200,000 to renovate and modernize equipment and buildings, plus additional funds as required for a large-scale exploration and development program.

The executive committee had reservations, fearing that the ore in sight would not justify such an expenditure, but they yielded to Starr's arguments as witnessed by the following letter.

San Francisco, May 9, 1898

George W. Starr, Esq., The Empire Mine, Grass Valley

Dear Sir:

I have the pleasure to inform you that the executive committee have accepted your ideas and that you are this day appointed Managing Director by the Board of Directors. You will be given a free hand in the equipment and development of our property with the understanding that if necessary to carry out your plans the sum of $200,000 will be furnished you, and that on completion of the plant the necessary expenditures for 600 feet per month of development, consisting of shaft-sinking and drives will be allowed. We sincerely trust that including the costs of the above development you will succeed in bringing all costs down to $7 per ton.

Very truly yours,
(signed) W.B. Bourn, President

Starr met the challenge with his customary energy, enthusiasm, and forcefulness. In July 1898 the work of pulling down and reconstructing was commenced, and all immediately necessary working parts were completed on January 7, 1899. Valuable new ore bodies were located exceeding all expectations. In George Starr's own words:

It is interesting to note that this re-creation was accomplished without hanging up more than ten stamps at a time, and the change from the tracks of the old incline and hoist to the tracks of the new 30 lb. rails in shaft and on the headgear was made in forty-two hours, during which forty stamps were dropping. I had no occasion to call upon the company for expenditures, but had the pleasure of producing sufficient to not only pay for all improvements, but to give the shareholders a dividend.

The new plant was completed in 1899 and underground development was rushed. The new 40-stamp mill was in continuous operation, crushing 30,260 tons in 1901.

The mill was of the usual type: Homestake pattern mortars, with five stamps per battery, each stamp weighing 850 pounds with a 7-inch drop at a frequency of 100 drops per minute, on a sequence of drops 1-5-2-4-3. The character of the ore was believed to require fine crushing, so 40-mesh screens were used.

The amalgamation plates were 4 by 17 feet in size, with a grade of 1½ inches to the foot. The amalgam cleanup pan (Berdan pan) consisted of a large cast-iron bowl, rotating 20 times a minute and set on an angle of 45°. Two cannon balls were placed in the bowl, which rolled in the opposite direction to the pan, thereby grinding and cleaning thoroughly all amalgam, headings, magnet iron, or any other contaminating material from the mill requiring grinding. A continuous flow of water washed out all waste material, leaving a high-grade gold amalgam.

Berdan pan.

Photo Credit: Hollis DeVines

This new plant gave consistently higher production yields, as indicated in the following tabulation (from G.W. Starr, 1900):

THE MILL GOLD RECOVERY	
By battery amalgamation	51.8 percent
Outside plates	23.1 percent
Vanners recovering sulphides	14.7 percent
Settling tanks for overflow, cleanup escapes, float sulphurets, etc.	1.2 percent
Slime plant	3.8 percent
Total extraction	94.6 percent
Milling costs per ton of ore crushed	$0.4924

Under Starr's efficient management, and with his superb experience gained in South African deep-shaft mining, the Empire entered years of profitable operation. With Starr as managing director, the Empire became a model mine in California in the early 1900s, attracting engineers from around the world.

Empire revenues now contributed heavily towards other than mining investments for the company, and it was in this prosperous era that large investments were made in the Spring Valley Water

Headed for work.
Man cars at surface
of Empire Mine.

Company, which became so vital to the growth of San Francisco. Another fine investment was in Fireman's Fund Insurance Company.

Starr loved Grass Valley, and it was due to his civic-minded efforts that mine property was given to the city for what is now Memorial Park and for its Olympic-sized swimming pool.

Throughout these years, Bourn remained close to Empire problems, but, with George Starr to carry the immediate burden of administration at Grass Valley, he found time for other projects, for extensive foreign travel with his family, and for participation in San Francisco community affairs.

In 1910 Bourn purchased Muckross House, a large Elizabethan mansion with thousands of surrounding acres, in County Kerry, Ireland, as a gift to his daughter, Maud, and her husband, Arthur Vincent. Five years later he built a large, modified Georgian-English home, "Filoli," near Woodside, California, about 25 miles south of San Francisco.

It is fitting that three of the Bourn family estates are now historic parks: Filoli, Muckross in Ireland, and the Empire Cottage and grounds in Grass Valley.

North Star Mine Historical Background

THE STORY OF THE EMPIRE MINE cannot be complete without including some historical facts about the North Star, which eventually became part of the Empire Star complex. Discovered in 1851, shortly after Gold Hill, its early history is remarkably similar to that of the Empire.

It was worked profitably for many years under several owners, just as the Empire. In 1860 it was known as the "North Star Quartz Mining Company," and in 1867 it was sold and reincorporated under the name "North Star Gold Mining Company." In 1868 a rich vein was struck and by 1875 the North Star shaft had reached a depth of 1200 feet on the incline. At this point the ore decreased in value until it could no longer be worked profitably and operations were discontinued.

As previously noted, W.B. Bourn in 1884 bought the idle North Star from its San Francisco owners. The mine was dewatered, a new stamp mill erected, and the mine became a good producer. It was sold to J.D. Hague and Associates in 1887, and the company was reorganized in 1889 as the North Star Mines Company. Under the guidance of Hague and mine manager, E.R. Abadie and later R.R. Roberts, the company prospered. Hague acquired many of the famous early day mines surrounding the North Star, including the Gold Hill, New York Hill, and Massachusetts Hill mines.

In 1895 the noted mining engineer Arthur D. Foote was brought to the North Star as superintendent and the mine became a second model mine in the community, luring young engineers to Grass Valley to spend a week or two studying its operations.

In 1901 a new vertical shaft, begun in 1897, cut the North Star vein at the vertical depth of 1630 feet, or about 4000 feet on the dip of the vein. This was named the North Star "Central" shaft, and the original shaft started in the 1850s was called the "Old North Star" or the North Star "Slant" shaft. From 1901 to 1902 profitable operations continued to the east and west.

The Central shaft was sunk in 1927 to the 8600 level, or 3700 feet vertically. A winze was sunk from the 8600-foot level to a final depth

FIGURE 1

CROSS SECTION THROUGH THE EMPIRE AND PENNSYLVANIA MINES, AFTER CLARK (1970).

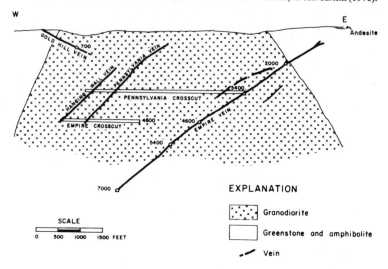

FIGURE 2

SECTION THROUGH THE NORTH STAR MINE CENTRAL SHAFT AND NO. 1 WINZE, AFTER JOHNSTON (1940).

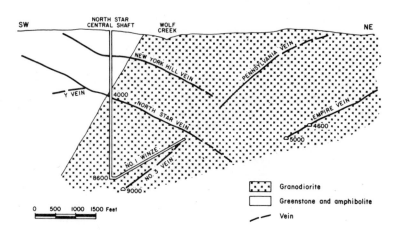

Cross sections from GOLD DISTRICTS OF CALIFORNIA by William B. Clark, California Division of Mines and Geology Bulletin 193, p. 53-80.

of 11,007 feet on the incline, almost a mile vertically below the surface.

The North Star and associated veins form a system of east-striking, north-dipping veins, as distinct from the Empire-Pennsylvania-Osborne Hill and Omaha-Allison Ranch systems of north-striking, west-dipping fissures and veins. With this orientation of veins and the proximity of the claims (see figures 1 and 2), conflict between these two giants, the Empire and the North Star, was inevitable.

Pennsylvania Mine Acquired by Empire

TWO OF EMPIRE'S NEIGHBORS, the Pennsylvania Mining Company and the W.Y.O.D. Company (Work Your Own Diggings), got involved in a monumental vein apex dispute which was settled in court in 1902, after 12 years of legal battles. The suit drained the resources of both companies, and in 1911 W.B. Bourn, Jr., got an option on both the properties in order to acquire the Pennsylvania Mine, which was equipped with a 20-stamp mill and a cyanide plant with a capacity of 28,000-30,000 tons per year. This proved to be an excellent investment; for example, in 1913 the Pennsylvania had ore production about equal to that of the Empire, with double the profits. In 1915 the Pennsylvania option was paid off and the property became an integral part of the Empire Mines and Investment Company.

In 1915 the underground crosscut from the 2600 Empire level cut to the Pennsylvania ledge, opening up valuable ore reserves. At this time the North Star Company shaft was sunk to about this same elevation, but the North Star vein either petered out or was badly faulted; however, the North Star did cut the Pennsylvania vein, including the deep ore that the Empire had found with its 4600 crosscut. The North Star Company acquired other properties, giving it some down-dip rights on the Pennsylvania vein at the point of discovery and forcing a compromise and adjustment between the two companies.

The Empire Cyanide Plant

THE EMPIRE MINES on December 1, 1910, put in operation a new cyanide plant, one of the most modern in the West. The plant had a capacity of 150 tons per 24 hours, somewhat higher than the 40-

stamp mill in operation at that time. The building was constructed of wood covered with corrugated iron; the framework on the inside was painted white and all the tanks and piping painted dark green. Electricity for lights and power was used throughout. The plant site on a hillside slope below the stamp mill permitted gravity flow of the pulp with minimum pumping.

The tailing from the sixteen 6-foot Frue vanners located in the stamp mill building flowed through 425 feet of 15-inch-diameter terra-cotta tile pipe (which was still in service 46 years later) into the Merrill classification equipment. The system comprised two primary Merrill settling cones 8 feet 4 inches in diameter having 50° sloping sides with the underflow distributed to five 4 feet 8 inch cones with sloping sides of 70°. Each of the five cones was fitted with a Merrill hydraulic sizer, dividing the pulp into sands and slimes. The slimes (fine material) free of sands from the seven cones and sizers were distributed to four 24-foot-diameter by 22-foot-deep clarifying (dewatering) thickeners fitted with false conical bottoms sloping 45° towards center.

The thickened slime was withdrawn continuously and fed by gravity to the first of four Pachuca air-agitated tanks in series, 10 feet by 18 feet deep, equipped with conical bottoms. The specific gravity of the thickened pulp was 1.4 to 1.6. Cyanide and lime solution was added to the Pachuca tanks with a cyanide consumption of 0.75 to 0.90 pound per ton. The pulp passed from number four tank to two

Sand leaching vats—cyanide plant—Empire Mine.

Oliver continuous drum filters, with the Oliver wet-vacuum pumps maintaining a vacuum of 22 inches. The filter cake was well washed with barren solution.

The sands from the final washing in the Merrill hydraulic sizers were fed through a Merrill gravity rotating distributor into one of four 155-ton capacity wooden leaching tanks equipped with overflow launders. Each vat held one day of stamp-mill production, thus allowing an average leaching time of 60 to 96 hours. After the vat was filled and drained, cyanide solution was percolated through the sand charge at a controlled flow rate which amounted to 144 tons of solution per 24 hours. Either barren solution or fresh water was used for the final wash of sand in the vat.

Each vat was fitted with a large gate valve and had a sloping bottom to facilitate sluicing out the sand tailings. The vat was equipped with a false bottom of timbers covered by canvas and slats covered with cocoa matting, permitting drainage of effluent solution. The false bottom was 8 inches high on the outside, tapering to 2 inches at the center gate. An automatic sluicing machine, similar to a lawn sprinkler, hanging from an overhead track was used for discharging or emptying the vats. It worked well, but used large quantities of water. Manual hosing was required for final sluicing.

The gold content of the slime and sand was about 45% and 55%, respectively. Gold extraction on the slime head, averaging about 0.10 ounce per ton, was 90 to 93%, and on the sands, with approximately the same gold content, extraction was 85 to 89%. Treatment costs, including the concentrates, refining, and assay office, for the year 1911 was $0.445 per ton of vanner tailing.

Precipitation of gold from rich, pregnant solution was by the well-known Merrill-Crowe zinc-dust method, which had replaced the use of metallic zinc shavings. Zinc dust consumption was 0.19 pound per ton of ore. The zinc-gold precipitate was collected in two filter presses with ten 36-inch frames and had a value of $18 to $30 per pound. The precipitate was fluxed with borax, soda, and silica and melted directly from the presses in tilting furnaces and cast into bars for shipment to the U.S. Mint in San Francisco.

The Frue vanner concentrate was reground in a 4-foot by 8-foot tube mill which handled 4 tons in 24 hours, grinding to a fineness of 85% minus 200-mesh. The value of the concentrate was $60 to $115 per ton.

The tube mill discharge was classified in a Merrill cone, with the overflow passing over 2-foot by 5-foot amalgamation plates on which 15% to 20% of the gold was recovered. The cone underflow was returned to the tube mill. The cone overflow was combined with the sands for vat leaching.

The highly successful Empire cyanide plant constructed in 1910, while not the first cyanide plant in California, was the most modern and efficient. The first cyanide plants were erected at the Crown Mine, New Zealand, in 1889 and on the Rand, South Africa, in 1890. The first plants in the United States appeared in 1891 at Mercur, Utah, and at the Utica Mine on the Mother Lode, California.

In the early 1900s, treatment of gold-bearing concentrates by chlorination gave higher gold recovery than by cyanidation, but was more difficult to operate. Had the Empire stamp mill tailings been impounded over the years instead of being sluiced down Wolf Creek, they could have been retreated by cyanide leaching at a substantial profit.

Many world-renowned metallurgists and equipment inventors contributed to the successful design and operation of the Empire plant. This list included J.V.N. Dorr, C.W. Merrill, Thomas Crowe, and E.L. Oliver.

Litigation, Plant Expansion, and Remodeling— 1913-1929

BY 1914 THE EMPIRE SHAFT reached the 4600-foot level on the incline. At that point, further sinking on this shaft was stopped as a result of North Star versus Empire litigation. The vein being followed steepened near that level so that the Empire shaft was going off under land to which the North Star Mines Company had agricultural title. The North Star brought suit against the Empire for invasion of territory and extraction of ore. The suit was settled out of court, and an arbitrary vertical boundary was established between these two adjacent properties. Further work in the Empire was then carried on by means of a winze from the 4600 level.

The Empire Mine during 1913 and 1914 launched a very extensive rehabilitation program, equipping milling facilities with 20 additional stamps and remodeling the existing 40-stamp mill, thereby increasing the crushing capacity from 47,000 tons per year in 1912 to 85,000 tons in 1915. The cyanide plant was increased in capacity to treat the higher tonnage of stamp mill tailings and also for treating the increased tonnage of Frue vanner sulfide concentrates. The new 60-stamp mill was equipped with 1575-pound stamps capable of

Wilfley concentrating table— North Star Museum.

Photo Credit: Hollis DeVines

crushing 5 tons of ore per stamp per day, which was the ultimate capacity until coarser battery screens were used at a later date. The replacement of the Frue vanners in 1916 with Wilfley concentrating tables resulted in more efficient sulfide recovery, and the 40-mesh screens on the stamp batteries were replaced with coarser 28-mesh screens.

In 1916 the Empire cyanide plant treated 108,000 tons of sands and slimes and 2400 tons of Empire and Pennsylvania concentrates. This tonnage was in addition to the Pennsylvania cyanide plant, which treated 28,000 tons. The Empire plant had a capacity to treat 350 tons per day of stamp mill vanner tailings. A Dorr thickener was installed in the cyanide plant and two new Rockwell furnaces for melting precipitates were added to the refinery. An electric railroad for hauling ore from the Pennsylvania shaft to the Empire mill and also waste rock from the Empire shaft was under construction.

In 1917 the California Debris Commission issued a requirement for the impounding of mill tailings, which had in former years been dumped into Wolf Creek. The Empire then made plans to construct a 63-foot-high earth dam, 835 feet long.

A further 20-stamp mill addition was completed in 1919, making a total of 80 stamps, which remained the ultimate number. To match the stamp mill capacity, the cyanide plant was enlarged by additional sand leaching vats, slime Pachuca agitators, Dorr thickener, Oliver filter, and one Devereaux agitator (for concentrates) with the necessary auxiliary equipment. A capacity of 400 tons per day became the maximum.

After treatment all tailings were impounded in the dam. The electric railroad, for ore haulage, from the Pennsylvania shaft to the Empire was completed and in operation; this allowed the stamp mill

Photo Credit: Hollis DeVines

Two Rockwell furnaces—Used to refine the gold/silver precipitate from the cyanide operation.

Photo Credit: Hollis DeVines

Bullion furnace—Gold from the amalgamation process or gold from the cyanide process was smelted for the final time in this furnace and poured into bars for delivery to the U.S. Mint.

and cyanide plant at the Pennsylvania to be shut down in favor of one central milling and plant facility. The Empire mill treated 142,000 tons in 1920.

The total cost of the Empire expansion was: stamp mill, $120,000; cyanide plant, $127,300; surface plant and equipment, $594,000. An additional $135,000 was spent on the Cottage and grounds.

In September 1916 the wages of miners were $3.00 per day and shovelers $2.25. By June 1920, miners were paid $5.50 and shovelers $4.25 per shift.

The following flowchart shows the processing method used at the Empire in 1929:

EMPIRE MINE ORE PROCESSING FLOWSHEET
1929

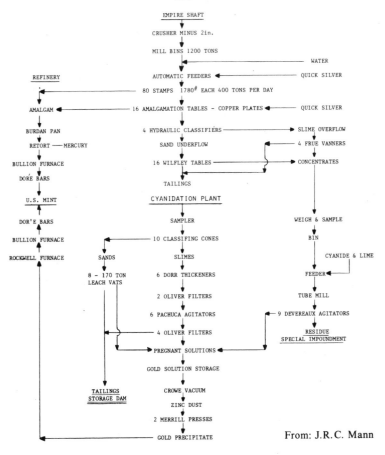

From: J.R.C. Mann

9

END OF AN ERA

FROM THE TIME THAT GEORGE W. STARR became managing director, with F.W. Nobs serving as general manager, the Empire Mine flourished in profit until the 1920s. At that point the effect of World War I on labor and material costs began to have a serious impact on mining profits.

Twenty-Eighth Annual Report

THE TWENTY-EIGHTH ANNUAL REPORT of the Empire Mines and Investment Company is reproduced in part on the following pages. Included are William Bourn's annual message to the shareholders of the company, the profit and loss statements for 1927, and statistical summaries of the mine's operation from July 1, 1900, to December 31, 1927.

A further summarization of these statistics permits derivation of the following data:

INCOME AND PRODUCTION DATA—1900 TO 1927

Bullion produced at $20.67 per ounce	$28,133,390
Operating Expense	14,670,295
Mining Profit	13,463,094
Revenue from Investments	10,545,906
Total Gain	$24,009,000
Production-Empire & Penna. Mines-Tons Ore	2,404,741
at a value of $11.60 or 0.56 oz./ton	$28,133,390
Operating & Development Expense, Taxes, etc.	$14,670,296
at a cost of $6.10 or 0.30 oz./ton	
Total Production was approximately 1.4 million ounces.	

TWENTY-EIGHTH ANNUAL REPORT

OF THE

EMPIRE MINES AND INVESTMENT COMPANY

THE EMPIRE MINES

FOR THE FISCAL YEAR ENDED

DECEMBER 31, 1927

Empire Mines and Investment Company

Capital, December 31, 1927 - - $25,000,000

*500,000 Shares of Stock, par value $50 each,
of which 400,000 shares have been issued*

Incorporated September 18, 1899

OFFICERS AND DIRECTORS

WILLIAM B. BOURN - - - - - - President

S. P. EASTMAN - - - Vice-President and Manager

E. L. EYRE - - - - - - - Vice-President

A. H. PAYSON   GEORGE W. STARR

MAUD BOURN VINCENT   WILLIAM H. CROCKER

E. J. McCUTCHEN   MILLEN GRIFFITH

GEORGE W. STARR - - Managing Director of Mines

F. W. NOBS - - - - - - General Manager

F. S. MARKEY - - - - - - - Secretary

San Francisco, Calif., February 1, 1928.

To the Shareholders of the
Empire Mines and Investment Company:

After years of operations our Empire Mines seem "tired" and the outlook is not hopeful, but work carried on incident to the Sultana option (adjoining Empire) has resulted in encouraging developments. Work has not been carried far enough to determine finally the ownership or value of the veins which have been opened up.

During the last ten or fifteen years conditions in the gold quartz mine industry have greatly changed. All the costs of producing an ounce of gold have increased, while the value of our product remains the same, or, stated in a different way, the value of gold has gone down due to the increase in producing costs.

Our investments are very satisfactory. Notwithstanding a raise in the "book value" of our Spring Valley Water Company stock and Fireman's Fund Insurance Company stock to one hundred dollars per share each, the market value of our stocks and bonds on January 3, 1928, was more than a million dollars over book value.

Respectfully,

W. B. Bourn,
President.

PROFIT AND LOSS FOR 1927.

DEBIT.

Office and General Expense		10,956.86
Legal Expense		4,204.36
Administration		11,271.05
General Welfare		2,000.00

Taxes:

Income, 1927	$37,560.51	
Franchise, license, etc.	5,096.93	42,657.44

Interest Paid	3,067.78
Contention Mines Lease	3,500.00
Depreciation — Mines Plant	34,601.53
Depletion of Mines	151,190.50
Compensation Insurance (to reserve)	14,255.53
	277,705.05

Net Profit to Surplus	736,735.86

	$1,014,440.91

SURPLUS ACCOUNT.

Cash Dividends paid in 1927	600,000.00
Surplus December 31, 1927	267,078.05
	$ 867,078.05

PROFIT AND LOSS FOR 1927.

CREDIT.

Empire Mines:

Bullion Production	$978,507.78	
Operating Expenses	845,293.38	
Gross Profit	133,214.40	
Sultana Option		
Cost of Development Work	69,043.73	$ 64,170.67

Bourn Block:

Rents		57,748.73	
Expenses	$28,789.11		
Depreciation	8,905.00	37,694.11	20,054.62

Revenue from Investments:

Dividends		505,422.50	
Bond Interest		192,442.73	
Other Interest		32,506.38	
Profit from Securities sold	263,028.79		
From which deduct			
Loss on notes	$23,115.00		
" " North			
Star stock	40,549.00	63,664.00 199,364.79	929,736.40

Rents and Royalties:

Rents	479.22
	$1,014,440.91

SURPLUS ACCOUNT.

Surplus January 31, 1927	130,342.19
Profit for year 1927	736,735.86
	$ 867,078.05

PROFIT AND LOSS
JULY 1, 1900 TO DECEMBER 31, 1927.

Expenses and Deductions July 1, 1900 to December 31, 1927.

Expenses:

Year	Administration and General Expense	Corporate Taxes	General Welfare	Other Expenses	Total
1900	$ 2,958.85				$ 2,958.85
1901	4,652.02				4,652.02
1902	2,936.49				2,936.49
1903	3,663.90				3,663.90
1904	5,285.60				5,285.60
1905	5,535.85	$ 20.00			5,555.85
1906	9,586.83	130.33			9,717.16
1907	10,753.08	450.00			11,203.08
1908	10,981.81	250.00			11,231.81
1909	7,025.37	250.00			7,275.37
1910	8,442.27	3,061.33			11,503.60
1911	7,854.20	3,214.69	$ 5,017.40		16,086.29
1912	6,303.22	3,826.61	7,500.00		17,629.83
1913	6,269.88	2,851.23	7,500.00		16,621.11
1914	6,330.36	9,186.01	20,177.13		35,693.50
1915	7,409.14	12,395.46	8,824.81		28,629.41
1916	14,596.05	24,119.15	26,248.67		64,963.87
1917	23,112.02	69,628.41	28,895.76		121,636.19
1918	33,537.62	128,014.59	5,416.67	$ 3,500.00	170,468.88
1919	26,696.53	85,326.91	4,027.00	3,500.00	119,550.44
1920	22,140.55	69,440.86	11,524.50	3,500.00	106,605.91
1921	22,396.82	43,575.58	4,950.00	3,500.00	74,422.40
1922	32,504.24	79,863.95	3,600.00	3,500.00	119,468.19
1923	66,739.62	51,033.67	550.00	3,500.00	121,823.29
1924	50,807.35	46,577.15	1,365.97	4,682.15	103,432.62
1925	55,020.60	48,348.72	5,825.00	12,356.06	121,550.38
1926	40,863.87	27,313.33	3,946.10	23,362.44	95,485.74
1927	26,432.27	42,657.44	2,000.00	6,567.78	77,657.49
	$520,836.41	$751,535.42	$147,369.01	$67,968.43	$1,487,709.27

Other Deductions

Mining:

Depletion of mines	$2,907,392.19	
Depreciation of plants	787,783.99	
Compensation insurance reserve	93,130.75	
Sultana Option — development work	274,331.67	$ 4,062,638.60

Pennsylvania — North Star Litigation ... 51,431.84

Losses on Mines:

Empire West Mines Company	275,951.51	
Northern Water and Power Company	434.994.11	710,945.62

Discount on Stock Sales:

9,000 shares @ $2.00 per share ... 18,000.00

	6,330,725.33
Profit carried to Surplus Account	17,889,274.46
	$24,219.999.79

PROFIT AND LOSS
JULY 1, 1900, TO DECEMBER 31, 1927.

Income July 1, 1900 to December 31, 1927.

| | Empire Mines | | | Investments | |
Year	Bullion Product	Operating Expense	Mining Profit	Revenue	Total
1900	$ 202,254.23	$ 87,112.82	$ 115,141.41	$ 939.17	$ 116,080.58
1901	720,214.73	142,751.77	577,462.96	29,285.06	606,748.02
1902	616,948.67	166,466.36	450,482.31	54,382.00	504,864.31
1903	493,620.48	170,947.02	322,673.46	57,377.68	380,051.14
1904	550,364.09	165,790.01	384,574.08	82,937.28	467,511.36
1905	553,583.49	166,112.41	387,471.08	511,138.25	898,609.33
1906	386,025.64	173,803.96	212,221.68	479,933.15	692,154.83
1907	321,840.75	185,315.79	136,524.96	125,733.90	262,258.86
1908	478,451.48	197,488.06	280,963.42	167,021.90	447,985.32
1909	508,218.68	223,968.49	284,250.19	209,574.47	493,824.66
1910	621,397.19	235,560.29	385,836.90	194,124.88	579,961.78
1911	613,865.91	339,776.22	274,089.69	213,409.19	487,498.88
1912	688,093.26	440,349.08	247,744.18	178,664.78	426,408.96
1913	1,117,786.47	415,459.83	702,326.64	200,165.16	902,491.80
1914	1,088,250.23	422,962.49	665,287.74	250,632.94	915,920.68
1915	1,471,583.06	494,342.80	977,240.26	287,574.23	1,264,814.49
1916	1,822,939.20	629,818.90	1,193,120.30	338,353.59	1,531,473.89
1917	1,848,623.62	744,384.62	1,104,239.00	455,200.05	1,559,439.05
1918	1,806,016.85	787,044.97	1,018,971.88	423,862.36	1,442,834.24
1919	1,639,314.96	926,960.64	712,354.32	535,528.59	1,247,882.91
1920	1,697,929.47	993,750.47	704,179.00	562,826.21	1,267,005.21
1921	1,325,954.57	967,679.18	358,275.39	586,115.49	944,390.88
1922	1,523,123.09	926,852.11	596,270.98	620,955.65	1,217,226.63
1923	1,260,089.21	971,708.15	288,381.06	640,609.74	928,990.80
1924	1,348,787.02	1,036,797.16	311,989.86	757,705.95	1,069,695.81
1925	1,232,465.20	893,757.43	338,707.77	796,567.65	1,135,275.42
1926	1,217,140.34	918,041.92	299,098.42	835,016.89	1,134,115.31
1927	978,507.78	845,293.38	133,214.40	950,270.24	1,083,484.64
	$28,133,389.67	$14,670,296.33	$13,463,093.34	$10,545,906.45	$24,008,999.79

Received July 1900 from Original Empire Mill and Mining Company........ 211,000.00

$24,219,999.79

Sale of the Empire

IN THE LAST 10 YEARS the mining profit declined from 1 million dollars a year to just over $100,000. Production costs per ounce of gold increased steadily on an upward curve from a production cost per ounce of $7.83 in 1910 to $18.04 in 1927.

Development work continued during 1927. One notable accomplishment was the sinking of the main shaft to the 7000-foot level, 928 feet below sea level. Managing director George W. Starr and general manager F.W. Nobs in their section of the annual report noted that on October 8 all mining operations at the Pennsylvania shaft were stopped. The shaft was to remain in good repair in order to pump to the surface the heavy inflow of water which otherwise would flood the Empire Mine. In order that the self-sustaining life of the Empire be prolonged, milling was reduced to 40 stamps in October when the Pennsylvania was shut down.

Development on the 3000 and 3800 levels of the Empire South shaft showed considerable encouragement, and indications in the bottom of the mine on the 7000-foot level were more encouraging than any findings since the 5400 level. However, with the higher operating costs and fixed hundred-year price of $20.67 per ounce of gold, the Empire's life appeared about ended when, in the following year of 1928, underground conditions did not improve.

In 1924 an option purchase agreement was made by the Empire management for a large number of claims known as the Sultana Group, which included the Prescott Hill, Orleans, Osborne Hill, and other mines to the south of the Empire. These mines had all been producers before 1900. Work was done from the Empire and Orleans shafts in an effort to locate the Houston vein system, but was unsuccessful.

During the year, mine development from levels off the Empire shaft totaled 5573 feet, of which 5020 was to the south in search of ore on the downward continuation of the Sultana vein system. Again, no promising bodies of ore were found.

The Empire also owned the Omaha, Lone Jack, Wisconsin, and Homeward Bound mines. These mines, south of the North Star, were small producers and inaccessible from the Empire workings.

Bourn suffered a stroke in 1922 that reduced his mobility to a wheelchair. In 1929 his daughter Maud died and this event, coupled with his own ill health and that of his wife, brought his lifelong interest in mining to an abrupt halt.

Over the years the Empire Mine had occasional periods when the veins appeared to be pinching out, but Bourn's faith never wavered. He employed the best mining talent available, and under his ownership the Empire became symbolic of state-of-the-art mining technology. However, the combination of ill health, age, depression over the loss of a beloved and only child, and the rapidly decreasing profits in mining all led to his decision to sell the mine.

By the end of 1929 he had sold the Empire to the Newmont Mining Corporation of New York. It marked the end of 52 years of family ownership of the Empire. Bourn died 7 years later, in 1936, at Filoli. George Starr, who retired to San Francisco after the mine was sold, died in 1940.

The production history of the Empire Mine for the years from its inception in 1850 to its sale to Newmont in 1929 is shown in the following tabulation:

TOTAL PRODUCTION OF THE EMPRIE MINE, 1850 TO 1929

Years	Tons Ore Crushed	Ounces Per Ton	Production
1850 to May 1854[1]	4,500	2.41	$ 225,000
1854 - 1863[2]	28,100	1.82	1,056,234
1864 - Fire burned Mill	—	—	—
1865 - 1878[2]	36,000 (?)	(?)	1,911,081
1879 - 1890[2]	59,508	1.88	1,713,840
1891 - 1928[2]	2,621,935	0.56	30,367,597
	2,750,043	0.62	$35,273,752

[1]Estimated from Browne and Taylor (1886).
[2]Johnston (1940).

10

NEWMONT AND THE EMPIRE STAR MINING COMPANY

BECAUSE THE NEWMONT MINING CORPORATION operated the Empire mine for 27 years, until the permanent closing in 1956, the story of Newmont is relevant, particularly during the late 1920s and 1930s, depression years when the revenue from the Empire kept Newmont solvent.

Colonel William Boyce Thompson, a mining entrepreneur, formed Newmont in 1921. It went public in 1925. The name was coined by combining parts of the words "New York" and "Montana" and was chosen because Thompson promoted mines on Wall Street and grew up in Montana mining camps. He brought into the company some of the foremost engineers in the world, with Fred Searls, Jr., as the leader and president.

Brief Background of Fred Searls, Jr.

FRED SEARLS, JR., was descended from pioneers who headed west from New York State in 1849 to settle in Nevada City. Fred was born in Nevada City in 1888 and grew up steeped in the local mining atmosphere, working in the Champion Mine while studying mining geology at the University of California, Berkeley. An honor graduate from Berkeley in 1906, it is interesting to note that Searls' thesis in fulfillment of requirements for a degree in geology was titled "Economics of Recycling the Empire Mine Waste Rock."

After graduation one of his early jobs was at the Goldfield Consolidated Mines, Nevada, where his association and friendship with George Wingfield of Nevada fame and Bernard Baruch, advisor to several presidents, lasted throughout their lifetimes.

Searls' work carried him around the world, but he had an

ingrained love for the Nevada County district and had always harbored the idea of operating one of the local gold mines.

In 1917, while exploring the geology of some of China's more remote provinces, the war news reached him. He returned to Nevada City immediately, and, with his boyhood friends, Billy Simkins and A.F. Duggleby, enlisted in the engineering regiment of the 1st U.S. Army Division, serving in France where he rose from private to lieutenant by war's end.

During World War II, Searls served in Washington helping to organize the War Production Board. In the Korean War he sent Newmont engineers to organize the Raw Materials Division of the Atomic Energy Commission. The author of this history was one of those engineers.

As head of Newmont, Fred Searls, Jr., made it into a multinational mining enterprise with operations on most of the world's continents.

Newmont Acquires the Empire Mine

FRED SEARLS AND W.B. BOURN had been friends for many years, and when Bourn decided to divest himself of the Empire in 1929 he thought first of Fred Searls and Newmont. Bourn's offer to sell the Empire for $250,000, a sum less than the insured value of the Empire stamp mill, was an offer that could not be rejected.

Searls wrote a personal letter to Colonel Thompson and described his assessment of the values which would accrue to Newmont. He also suggested that, if Newmont was not interested, Searls and others in the Nevada City-Grass Valley area would be. A deal was struck and Newmont Mining Corporation became the new owner of the Empire Mine.

Newmont Acquires the North Star Mine

CONCURRENTLY WITH THESE NEGOTIATIONS, the owners of the North Star Mines Company approached Fred Searls, Jr., with a plan that they put the North Star Mine plus $150,000 working capital into a new company, in return for which they would receive 49% of the new shares.

Several factors made the North Star property attractive to Newmont, the new owners of the Empire. In the North Star Mine one of the best ore shoots was continuing into the Omaha claims owned by

Newmont's Empire. Two of the Empire veins appeared likely to pass from Empire-owned mining claims into North Star ground. Other unmined ore and undeveloped areas existed near the common boundary or in places where extra-lateral rights were in dispute. Also, the main Empire ore shoots merited further exploration in depth, which could be done more efficiently from the North Star workings. The offer from the North Star owners was, therefore, accepted, thus removing all potential legal problems.

North Star Production Data

The production history of the North Star from its inception to its acquisition by Newmont is shown in the following table:

YEARS	TONS ORE CRUSHED	PRODUCTION IN DOLLARS	PRODUCTION IN OUNCES	ORE GRADE OZ. PER TON
1851—1857[1]		250,000	12,500	2.00
1860—1874[2]		2,500,000	125,000	1.00
1884—1928[3]	2,630,780	$29,032,155	1,404,555	0.53

Estimated total production $31,782,155

[1] Estimated by Lindgren (1896).
[2] Annual Reports of North Star Mines Company.
[3] Johnston, Jr., 1940.

On May 1, 1929, Newmont acquired 51% of the issued stock of the North Star Mines Company. On May 21, 1929, the Empire Mines and Investment Company and the North Star Mines Company were merged into a new company, the Empire Star Mines Company, Ltd., and the properties and facilities of both former companies were conveyed to the new one. The merger brought over 3700 acres of mineral rights under single management.

11

THE EMPIRE STAR
UNDER NEWMONT CONTROL

THE MERGER CREATED A CONSOLIDATION of several operating mines in the Grass Valley district. Among those included were the Empire, Pennsylvania, Sultana Group, Omaha Group, North Star, and adjoining mines. The table below presents the production history of the member mines prior to their merger in 1929.

	GOLD PRODUCTION OF MINES PRIOR TO MERGER PRODUCTION - 1851 TO 1929		
MINE	TONNAGE	AVE. GRADE OZ./TON	TOTAL OZS. GOLD
Empire	2,750,000	0.61	1,677,500
Pennsylvania & W.Y.O.D.	365,500	0.54	134,000
Orleans, Sultana, Osborne Group	400,000	0.91	364,000
North Star	2,724,000	0.55	1,498,200
New York Hill	100,00	0.73	73,000
Gold Hill	190,000	1.00	190,000
Massachusetts Hill	208,000	0.92	191,360
Cincinnati Hill	2,000	0.40	8,000
Omaha, Lone Jack, Home-ward Bound, & Wisconsin Group	220,000	0.70	154,000
Total Empire Star Mines	6,950,500	0.62	4,290,060
Value of Production 1851 to 1929	$90,000,000		

Fred W. Nobs, manager of the Empire Mines under George Starr, was made general manager of the Empire Star Mines Company, Ltd. Nobs reported that the consolidation of the two mining properties opened up vast new promising exploration prospects. Between May and December 1929, work on the details of the merger went forward. By the end of the year all aspects were completed, including reorganization of the management structure and the reassignment of personnel. During this period the Empire 80-stamp mill crushed only 89,000 tons, with a $19,000 loss.

By 1930 the two separate operations were consolidated under a single staff and the Empire stamp mill and cyanide plant operated at full capacity; also, 50 of the 60 stamps in the North Star mill were put into operation. This substantial increase in tonnage milled clearly pointed out the considerable advantage of a unified management for both properties. Another innovation was the installation of the flotation process in the North Star mill for treatment of lessors' ore. This was one of the early applications of the flotation process to a gold ore. It proved quite successful and later became standard practice in many gold operations. An aerial tramway was constructed for transporting ore between the North Star and Empire to allow more flexibility in milling North Star ores.

A profit of $86,675 was made for the year, which was quite low for the high tonnage milled due to treating reserves of low-grade material. However, drifting and raising underground on the newly discovered "Newmont Vein" in the Empire developed considerable tonnage of good grade ore.

Acquisition of the Murchie Mine

ACQUISITION IN 1931 by the Empire Star of the Murchie Mine, 2 miles east of Nevada City, was concluded. The Murchie was a relatively shallow ore deposit. Although the mine was discovered in 1878, the high sulfide content of the ore was not amenable to known metallurgical methods except by smelting until the use of the flotation processes of the 1930s. This ore yielded a recovery of 95% by flotation. The concentrates were amenable to treatment in the Empire concentrate cyanidation circuit with a 90% recovery of the gold contained in the sulfides.

Mineral rights comprised 432 acres. The shaft was reopened and workings dewatered in 1928. The vein average width was 4 feet, consisting of quartz carrying free gold with 2.5% sulfides assaying about $200 per ton in gold and silver.

During the year, the Empire Star underground development reached an all time high of almost 20,000 feet, a fine accomplishment. A total of 214,000 tons was milled for a recovery of $9.44 per ton, equal to 95.9% of the gold in the ore. Due to the successful financial year, an initial dividend of $1 per share on 108,700 shares was declared.

The development of the mines, including the Murchie, was in full swing for the next 2 years. Because of the Great Depression, labor, including skilled miners, was plentiful with "rustling lines" (men looking for work) comprising 25 to 40 men each morning at 7 o'clock. Gold mining was one of the few operating industries during the early and mid 1930s. The Empire Star, under the guidance of Fred Searls, Jr., continued the policy of acquiring other gold mines. The most recent was the Northern Empire Mine, Geralton, Ontario, Canada.

Operating the Grass Valley mines proceeded as scheduled, although the possible exhaustion of the North Star and Pennsylvania properties caused considerable concern. The ore reserves left by the former management were soon depleted and development of new ore bodies was unpredictable. The tonnage of ore required for profitable milling from the narrow veins was difficult to maintain. However, management was reluctant to close any mines and discharge several hundred men under depression conditions of high unemployment. Even with over 22,757 feet of underground development plus 1800 feet of diamond drilling, reserves for the 228,000 tons milled in 1932 were not replaced.

Federal Regulations Permit Higher Gold Prices

IN 1933 THE ENACTMENT OF FEDERAL REGULATIONS permitting the sale of gold at world prices, which were 40 to 50% higher than the domestic standard of $20.67 per ounce, was the salvation for the company operations. Wages and salaries of all employees were increased 11% and the possibility of closing the mines evaporated overnight.

The program of persistent underground development of 22,000

feet plus over 5000 feet of diamond drilling was rewarded by the discovery of a major ore body on the 8600 level of the North Star Mine.

The Murchie Mine was producing at an annual rate of around 80,000 tons of ore at a grade of 0.415 ounce of gold and 1.48 ounces of silver per ton. The mineral values concentrated by flotation with recoveries of 93.8% gold and 87.2% silver produced 1800 annual tons of flotation concentrate which were cyanided at the Empire plant, yielding an overall recovery of approximately 90%.

At the Northern Empire Mine in Ontario, Canada, a new railroad siding, post office, and mine buildings came into existence and became known as Empire, Canada. The total investment by Empire Star Mines in the Northern Empire facility was $258,752.

An interesting and rather humorous incident occurred as a result of a shipment of equipment from the Empire Mine to the Northern Empire Mine. Fred Searls, Jr., was convinced that the high gold recovery (95%) in the Grass Valley operations was due to the effective stamp milling which crushed the ore to minus 28-mesh, a particle size that resulted in nearly complete amalgamation. As a result he had the local management locate a 20-stamp mill at an abandoned gold mine in the nearby mountains. The mill was dismantled, crated, and shipped to Canada. However, the heavy iron parts comprising the stamp mill were held in custody for a long time at Port Arthur, the Canadian port of entry, because the Canadian customs manuals listed no such equipment used in Canada.

Acquisition of Zeibright Mine

AN EMPIRE STAR DEAL was made with Fred Searls, Jr., owner of the Zeibright Mine located in Bear Valley, 25 miles east of Grass Valley, for development and operation of this property as a potential low-grade, large-tonnage gold mine.

In the following 2 years Empire Star profits were high, mainly due to the higher selling price for gold. The new ore body discovered on the 8600 level of the North Star mine extended upward on dip length for 1100 feet. The aggregate tonnage of total mine ore reserves of 400,000 tons of 0.386 ounce gold per ton was the highest in the mine's history. Dividends paid out for the two years (1934-1935) for 110,620 outstanding shares was $18 per share. All subsidiary mines, Murchie, Northern Empire, and Zeibright, were in production. A new pros-

Diamond drill at North Star Museum. Used in exploration for new bodies of ore.

pect at Browns Valley, 24 miles west of Grass Valley, was under development.

In 1936 an effort was made to explore all reasonably attractive development possibilities while the mines were still in vigorous production; crosscutting, drifting, and raising amounted to 28,609 feet, and diamond drilling to 7072 feet. This work included considerable footage in the outlying Prescott Hill and Sultana mines south of the Empire and in the Omaha property adjoining the North Star to the south. Results were so discouraging that these workings were abandoned, and the extensive country to the south of the producing mines was no longer considered for potential prospecting.

The tonnage milled in 1937 from the Empire, Pennsylvania, and North Star Mines was 238,743 with a high gold recovery of 0.449 ounce per ton. Development included deeper sinking of the No. 5 winze from the North Star 8600-foot level to the 9800 level which was 4220 vertical feet below the collar of the Central shaft. Both the Murchie and Northern Empire mines operated profitably; however, the Browns Valley and Zeibright Mines were unprofitable.

The years 1937 and 1938 were marred by the labor unrest which swept through the country in the wake of the drive by John L. Lewis and his CIO to organize American labor from coast to coast. The Grass Valley-Nevada City area did not escape this turmoil. Several ugly confrontations occurred at the Murchie and at the Idaho-Maryland mines.

Outside organizers were viewed suspiciously by the local miners and the community generally. This stemmed, no doubt, from many sources, but the fact that the mines of Grass Valley and Nevada City were operating at high capacity and paying dividends while the rest of the nation was suffering from long unemployment lines was very probably a strong reason for the desire of the local citizens to maintain a status quo.

In 1938 engineers at the Empire Star recommended the undertaking of exploration at a depth of 1730 feet below sea level and considerably below the level of any existing workings.

Murchie Mine Closes Production, 1938

THE MURCHIE HAD A SERIOUS DROP in mine tonnage which, combined with a much lower gold and silver content, indicated that the veins had bottomed. Deeper development failed to find any commercial ore. The decision was made to pull out the mining equipment in preparation for allowing the mine to fill with water up to the 800-foot level. A 6000-foot-long crosscut was projected from the 400-foot shaft level to be driven in a southeasterly direction. The object was to prospect the North Banner area adjacent to the Lava Cap mine, an unexplored area. This district produced $175,000, principally from the Woodville claim, from quartz ore carrying free gold and 5% sulfides with a high silver content. From 1932 through 1938, the Murchie Mine, as part of the Empire Star group of mines, produced 533,000 tons, yielding a revenue of $5,106,000. Previous production was approximately $2,000,000.

During the Great Depression of the 1930s, gold mining was the most profitable mining industry. Newmont, having acquired the Empire Star group as well as other gold properties, had some 10 years of good income at a time of corporate need for funds. The combined production of the Empire Star Mines averaged between 600 and 700 tons of ore per day during the 1930s. The income to Newmont for the years 1931 to 1943 was $3,216,210. Another major advantage of the Empire Star acquisition was that it changed Newmont's status, for the first time, from that of a holding company to one actively engaged in mining operations. This was the forerunner of Newmont's success as a world class mining company, and in later years it was often stated that the sun never set on one of Newmont's worldwide operations.

Another gold mine in which Newmont had a highly profitable interest in the late 1930s was the famous Getchell mine near Winnemucca, Nevada, an intriguing story itself.

New Milling Method Installed

STAMP MILLING, PLATE AMALGAMATION, Wilfley tabling, cyanidation of tailings and concentrates for gold recovery as practiced for over 20 years was continued by Newmont for a few years until flotation was installed as a more efficient recovery method.

The innovative metallurgical flowsheet developed at the Empire and North Star was to use the stamps as secondary crushing units. They were retained due to the high cost of bypassing these efficient crushers, already installed and amortized. The flow scheme then adopted in 1936 comprised a jaw crusher, stamp mill crushing with 1/2-inch screens on the batteries, ball mill grinding to minus 48 mesh, hydro-pulsating jigs, and rake classifier with the overflow to flotation which produced a tailing assaying less than 0.01 ounce gold per ton. The jig hutch gold product was pumped to a Wilfley table totally enclosed in heavy wire netting for security where a free gold concentrate was continuously amalgamated, recovering 50 to 65% of the gold. The flotation concentrate assaying 4 to 6 ounces gold was sent to the cyanide plant for treatment.

The plant amalgam was retreated in a Berdan tilted, revolving cast-iron saucer-shaped pan 3 feet in diameter by 1.5 feet deep containing a cast steel ball about the size of a bowling ball. All contaminating material was washed over the edge of the Berdan pan, the amalgam compressed to remove excess quicksilver, retorted, and melted into gold bars of 1000 ounces Troy weight. The mercury from the "clean up plant" and from the refinery retorts was returned for reuse. Security practices were nonexistent and depended on the integrity of the employees.

The flotation concentrate, replacing the table concentrate delivered to the cyanide plant, was first ground in a tube mill (length 4 times diameter) in closed circuit with a specially developed bowl classifier to obtain a product size of 98% minus 400 mesh for efficient cyanidation which was done by batch agitation with a retention time of 120 hours. The cyanide residue assayed approximately 0.20 ounce gold per ton for a 95.96% recovery. The residue was stored separately. The only method at that time for recovering the contained gold would be by roasting or smelting.

The coarse stamp crushing, ball milling, and flotation occupied about 10% of the former floor space and eliminated sand-slime cyanidation in addition to having lower operating costs and higher gold recovery. This gold ore milling scheme soon became standard practice in many gold mines and is still used today in Canada.

The flow scheme used at the Empire in 1939 is shown below.

EMPIRE FLOWSHEET
1939

20 - STAMPS
REPLACE 28 - MESH BATTERY SCREENS WITH 1/2 in. OPENINGS

200 TONS PER DAY

36 in. PAN AMERICAN JIG ⟶ HUTCH PRODUCT

OVERFLOW   6 ft DORR CLASSIFIER   WILFLEY TABLE ⟶ TAILS
MINUS 65-MESH

FLOTATION   SANDS   GOLD STREAK

CONCENTRATES   6'x6' BALL MILL   AMALGAMATOR

TAILS DAM

CYANIDE PLANT   AMALGAM TO REFINERY

From: F.W. McQuiston, Jr.

Summary of Empire Star Mines, 1939 to 1941

IN THE NEXT TWO YEARS, 1939-1940, underground development in the Empire Star aggregated a record 54,000 feet (10 miles), which included work on the 10,600-foot level of the No. 5 oreshoot in the North Star that had a very satisfactory gold grade. The Pennsylvania shaft was equipped with a new head frame, and other surface equipment was modernized. A peculiar circumstance occurred when the mine workings were invaded from above by diamond drill holes drilled from the adjoining Golden Center Mine property into a remote section of the Pennsylvania. They were running their mine water into a Pennsylvania Mine sump for disposal to save pumping costs to the surface. This prompted a lawsuit that the Empire won.

The Murchie southeast crosscut to the North Banner Mine district, a length of 6000 feet, yielded about 10,000 tons of 0.65 ounce per ton gold ore with 2.50 ounces silver per ton. The Browns Valley

Mine won an extra-lateral rights lawsuit and operated profitably. The Northern Empire Mine became unpayable in depth.

In March 1940, during an exceptionally heavy downpour, several large sections of the tailings pile below the Zeibright mill, which was situated on the south bank of the Bear River, moved like a glacier into the river and were carried down into the PG and E powerhouse, effectively shutting down the hydroelectric power plant. The Zeibright development had reached the 1400-foot level below the No. 1 adit and indicated the existence of at least 2000 tons of ore per vertical foot of depth. Evidence that the ore grade would persist in the lower levels indicated that this mine could become a profitable, high-tonnage, low-cost producer, providing a tailing disposal site could be developed.

Banner Year for Production, 1941

1941 WAS A BANNER YEAR for the Empire Star with 248,300 tons milled from which 0.349 ounce of gold per ton was recovered. In spite of a shortage of skilled miners, 25,641 feet of development work was accomplished. Fred Nobs estimated ore reserves as of January 1, 1942, at 507,800 tons at an average grade of 0.38 ounce per ton. Ore not previously known to exist was found on the 7000-foot level of the Empire and on the 10,450 level in the North Star. The underground workings comprised over 370 miles of shafts, winzes, drifts, and crosscuts. The greatest depth of 11,007 feet on the incline was reached in the North Star shaft, almost 1 mile vertically below the surface. Open stope mining continued to be used throughout the mines.

Prospecting in the vicinity of the North Banner property continued from the Murchie shaft throughout the year. The Murchie property now having been exhausted was being stripped of equipment before being permanently abandoned.

The Zeibright Mine was allowed to fill with water to preserve the timbers. A start was made on the Omega Adit to the north which, if extended far enough, would connect with the Zeibright No. 1 Adit. The purpose of driving the two adits was to dispose of tailings into the old Omega hydraulic pit some 4 miles to the north. A gold-bearing placer deposit supposedly existed between the Omega and Zeibright on the course of the adit.

The Northern Empire Mine in Ontario, Canada, was closed and

abandoned; however, a satellite mine, the Magnet, continued to pay dividends.

Closure of Gold Mines for War Support

With the advent of war in 1942, the War Production Board, an instrument of the United States Government, issued on October 8, 1942, "Order L-208." After defining gold mines as nonessential industry to the war effort, L-208 ordered the gold mines to take immediate steps to close in the shortest time possible. The objective was to force the mine workers into shipyards and eliminate use of explosives, hoisting cable, and other supplies now essential to the war effort. The irony of the Order was that the United States manufacturers still supplied Canadian and South African gold mines.

This was the first time that gold production in the Grass Valley properties completely ceased for more than a few days during the 90 years since their initial operation began. In compliance with the order, the mills in Grass Valley ceased operations in November 1942, but the mines were kept unwatered and in repair. The ore reserve as of the date of closing was estimated at 536,000 tons at 0.372 ounce per ton.

The Browns Valley Mine, which had confined operations to the Dannebroge shaft, contained some good ore which warranted further development when circumstances permitted.

For 3 consecutive years, the mines and mills remained closed. However, continuation of the policy of investment of the working capital consisting of dividends, interest, and profit from the sale of securities offset the operating loss of the mines. The amount for 1944 was $253,000.

Gold Mines Reopen, June 1945

War production board order L-208, which by government fiat closed the gold mines in November 1942, was lifted on June 30, 1945. The acute shortage of miners and other underground labor due to government subsidies of the Premium Price Plan for nonferrous metal mines had advanced wages not tolerable in the gold mines. The Browns Valley Mine, which required only 50 men, was reopened on a trial basis but was unprofitable and had to be abandoned. This attempted operation proved the suspected difficulties of trying to operate under existing conditions. Assuming the price of gold to be $35 per ounce, it was necessary to either obtain, within a reasonable time, an adequate labor force at wages somewhat below current pay for similar skills or else close the mines, which would be equivalent to abandoning them. This was a difficult decision for the managers of the many California gold mines.

Persistence throughout 1946 of the unsatisfactory labor conditions prevented any significant production. Demands for wage increases were unacceptable to management because such increases would have brought about operating costs exceeding the value of the gold produced.

In 1947 the Grass Valley properties resumed partial operations and small-scale production by giving leases of underground blocks in the more remote part of the mines. Blocks of ground were assigned to individual groups of miners on a share basis in which the company provided all support services, including blasting powder, and shared with the lessors the production revenue. Gold sales for the year amounted to $600,000, resulting in a net loss of $232,000. The leasing program continued to expand and company mining and development operations were also conducted on a small scale. These operations increased reserves more rapidly than the ore was extracted. Lessors for the most part worked in ground not included in previous ore reserves. Value per ton of ore mined by lessors was higher than that formerly mined by the company, having an average content of 1.50 ounces of gold per ton. Total income for the year of 1948 showed a small profit.

Leasing and restructured company mining operations continued through the next 2 years, but production was limited by lack of skilled miners. Operations during 1951 were on a curtailed basis throughout the year, increasingly limited by available labor and higher costs. The deepest workings of the North Star were allowed to fill with water and work was also suspended in the Empire Mine below the 4600 level. Lessor ore averaged 1.95 ounces, and company ore 1.30 ounces per ton.

The noose was slowly tightening and it was only a question of time until the loss in revenue forced the closure of the 100-year-old mines.

The life of these great mines, the Empire, the North Star, and other mines comprising the group, was finally drawing to a close. They had served as a store of wealth for the Bourns, Newmont, and many shareholders and were the main support for the town of Grass Valley.

The gold production under the management of Newmont Mining Corporation with the guidance of Fred Searls, Jr., was as follows:

TOTAL ESTIMATED PRODUCTION—EMPIRE STAR
1929 THROUGH 1956

YEARS	TONS MILLED	RECOVERED OZ./TON	TOTAL OUNCES RECOVERED
1929-1942	3,130,766	.392	1,215,321
Nov. 1942 to June 30, 1945 (Operations suspended due to WPB Order L-208)			
1945-1956	265,000	1.320	350,000 (approx.)
Total	3,395,766		1,565,321
Total 1850-1956	±10,346,000	0.570	5,855,000

12

THE CLOSING OF
THE EMPIRE STAR MINES

IN THE NEWMONT MINING CORPORATION'S ANNUAL REPORT for the year 1956, the following statement was made with respect to the Empire Star Mines Company, Limited:

> During 1956, the Company milled 20,850 tons of gold ore mined from its Grass Valley Properties, from which a gold recovery valued at $36.22 per ton was realized. The Company sustained a net loss for the year of $256,517. Mining operations at Grass Valley were suspended on July 5, 1956 as a result of a strike by the local labor union. Wage demands cannot be met at the present price of gold, and recent mine development has been disappointing. Underground dams are being constructed preparatory to allowing the mines to fill with water. These dams will effect a separation of the mines and will facilitate their separate unwatering in the event operations are resumed in the future.

With no hope in the foreseeable future for better conditions, the gold mining industry throughout the state of California closed their mines, allowing them to fill with water and to cave in. The Empire and other mines in the district suffered the same fate. In due time all areas and equipment of the Empire Star were cleaned up, with particular attention to salvaging underground and surface equipment. Concrete foundations, particularly around the stamp mills, that could harbor quicksilver accumulations (which because of its high mobility and gravity moved quite freely until firmly trapped) were carefully cleaned up, and invariably carried gold. The cleanup of the stamp mill, concentrating table floor, cyanide plant, and particularly the 45-year-old concentrate pipe connecting the stamp mill with the cyanide plant yielded several thousand ounces of gold.

The plant equipment was disassembled and many buildings torn down. An auction was held September 25-26, 1959, to dispose of usable equipment, and the stamp mill was sold as scrap iron—a sad fate for such a historical producer.

At the time of the mine's closure, the Empire Star property comprised some 6000 acres of mineral rights with approximately 4000 acres of surface holdings in the Grass Valley-Nevada City district.

In the vicinity of the Empire shaft were many interesting stone, brick-trimmed buildings built around the turn of the century, such as the administration complex housing the general office, senior staff offices, warehouse, gold refinery, mine safety office, and other offices. Other buildings constructed of stone were the Bourn Mansion (called the Cottage), a group of stables, and a carriage house. The mine buildings of sheet iron were the hoist room, change room, machine shop, blacksmith shop, electric shop, foreman offices, stamp-concentration building, and the cyanide plant. The latter two buildings covered many thousands of square feet. The shaft head frame was erected of heavy bolted timbers supporting the sheave wheel and heavy cable for man skips handling about 40 men per trip. The man skips were removed when the ore skips were in use.

Over the years many important mining innovations, inventions, and improvements, both in equipment and in operating processes, were developed in the Grass Valley gold mines. Just a few of those which were either invented by men of the Empire and North Star or were developed through experimental use at the Empire Star Mines were:

The "Go-Devil" ore cars, for lowering ore from stopes to the level below.

The Oliver Rotary Drum Continuous Filter—a major contribution to the effectiveness of the cyanide process.

Dorr Thickeners and Classifiers—for hydroseparation of finely divided solids.

Delay Action Electric Exploders—eliminated dangerous and sometimes ineffective hand-cut fuses.

The "Torpedo" lightweight machine drill—highly efficient, it remained in use until the air hammer drill superseded piston drills.

The Grass Valley area was also noted for the skill and reliability of its miners, who were drawn from all nationalities and from all parts of the world. Many were of Cornish (Cornwall, Great Britain) descent. Known as "Cousin Jacks," they represented the very best in hard-rock miners with vast experience in working deep mines.

In the history of gold mining in California, the Empire Mine stands preeminent, not alone for its wealth, but for what the mine, above all others, has given in the way of well-applied endeavor—the pioneer in deep mining, and the first to regard mining a legitimate business. Behind that preeminence, and the fountainhead of that pioneering endeavor, are the notable geological, mining, and metallurgical engineers who contributed to the success of the North Star and Empire mines. A partial list would include Waldemar Lindgren, Herbert Hoover, John Hays Hammond, Fred Searls, Jr., George W. Starr, James D. Hague, Arthur D. Foote, Arthur B. Foote, Errol Mac Boyle, Fred Nobs, W.D. Johnston, J.R.C. Mann, Arthur Kendall, Robert Hendricks, William A. Simkins, and Frank McQuiston.

Although active mining ceased in 1956, closing a large mine is a slow, laborious task. To the small group of men, whose lifetime careers were such an integral part of the Empire workings, it was a somber activity. Two days before Christmas 1961, the remnants of what had once been a proud and powerful work force filed silently off the Empire premises for the final time. For well over five generations, these men and their predecessors had mined the precious gold from the grasp of the Sierras; now it was the finale to a way of life that would never return.

Any epitaph to the Empire Mine should, perhaps, include the following: Grass Valley, located in the foothills of the Sierra Nevada at an elevation of 2400 feet, is the oldest and best-known mining town in the United States. It flourished with the vigor of "Pioneer days" and was tempered by a growth of 135 years. During this time the Empire Mine operated continuously for 106 years, helping to make this the richest and most famous gold-mining district in California. Grass Valley, now a prosperous city, began its existence with the discovery of gold in quartz ledges in 1850. To the Empire Mine, Grass Valley owes a debt of lasting gratitude, for in the history of that mine, more than in all other circumstances combined, lies the origin and the impulse leading to Grass Valley's growth from a temporary mining camp to an attractive city of permanent homes.

Photo Credit: Hollis DeVines

Model of the Empire Star and associated complex of mines as it existed at the close of operations in 1956. Showing over 367 miles of drifts, crosscuts, and winzes, this model was once protected from all eyes but those of top management. It is now open on special tours to the public.

13

Empire Mine Purchased by California State Department of Parks and Recreation

After the Empire had been unutilized for over 12 years, the California State Department of Parks and Recreation became interested in the site as a possible state historic park. Negotiations between the State and Newmont representatives, including the author, to purchase the Empire commenced January 9, 1974. The Empire Mine buildings were located on a 125-acre parcel, which was a part of an 1122-acre tract. Final agreement between Newmont and the State of California was for the purchase of 777 acres, including all buildings. The agreement was finalized on December 14, 1974. The purchase price was $1,250,000. This transaction did not include mineral rights below a depth of 100 feet of the surface. The area was designated the "Empire Mine State Historic Park." Newmont retained a 47-acre tract in addition to other acreage of surface ground, with good roads, electric power, and water access for future mining and milling location should reopening of the Empire Star Mines be warranted.

The California State Department of Parks and Recreation is fully deserving of great credit for renovating and restoring the rundown, vandalized buildings and the Bourn grounds and gardens to the excellent conditions existing in the early 1900s.

ADDENDUM

Historic Gold Production of Grass Valley and Nevada City Mining Districts

THE PRODUCTION OF THE DISTRICTS is difficult to determine accurately because of the incompleteness of the early records. Lindgren (1896b) estimated $113 million production from 1849 to 1893, with at least 60% credited to lode mines. From 1903 through 1958, Nevada County produced 7,119,353 ounces of lode gold with almost all from the Grass Valley and Nevada City districts. Converting Lindgren's estimate to ounces, the total production of the district through 1959 was approximately 10,408,000 ounces of lode gold and 2,200,000 ounces of placer gold as calculated by the United States Geological Survey in Koschmann and Bergendahl's P.P. 610, 1968, "Principal Gold-Producing Districts of the United States." The U.S.G.S. does not include production from 1893 to 1903, which was about one-half million ounces.

The above production figures, shown in ounces, are no doubt on the low side because the production from the mines in the period from 1850 to 1890, recorded in dollars, was subsequently converted to ounces by using $20.67 price per ounce. There are many references showing that some miners and mines received only $15 to $18 per ounce. Also the grade of ore was considerably higher as the production of the early mining operations was reported as the dollar value for free gold only and did not include gold-bearing sulfide values.

The following table shows the gold production of the Grass Valley and Nevada City mining districts:

GOLD PRODUCTION OF GRASS VALLEY AND NEVADA CITY MINING DISTRICTS	MILLION $	MILLION OUNCES
Empire-North Star Group Mines	135.664	6.035
Other Mines in Grass Valley	13.880	0.694
Idaho Maryland Mine	64.241	2.295
Nevada City District	50.000	2.081
Total District Lode Gold	263.785	11.105
Total District Placer Gold Production		2.200
Total Nevada County Production		17.016

Historical Price of Gold

GOLD INCREASED ALMOST FOUR TIMES in value during the period 1344-1717. The dollar value of gold was first maintained at a fixed price by the 1792 Mint Act based on a silver/gold ratio of 15:1. This established a price for gold at $19.38 per ounce, which was maintained until 1834 when the silver/gold ratio was changed to 16:1 with a fixed gold price of $20.67 per ounce. This price persisted until 1861 when "the Civil War Era" caused prices to fall. In 1879 the government reestablished the price at $20.67. This price remained through the inflation of World War I and the depth of the Great Depression. The price of gold began to rise in April 1933 until it attained $30 per ounce in September. In 1934 the government devalued the dollar by increasing the price of gold to $35 per ounce, which was the first price increase above $20.67 in 142 years. The price was fixed by Presidential Proclamation on January 31, 1934, at $35 per ounce. (From James A. Anderson, Homestake Mining Company, January 1982).

It was this period of the 1930s that made the Empire Star such a fine investment for Newmont because labor became more abundant and efficient, while wage rates and the cost of supplies decreased. However, War Production Board Order L-208 suspended gold mining operations in the United States from 1942 to 1945, during World War II. Future gold price increases had no effect on the Empire because the mines were closed before the higher prices came into effect.

In the period of the 1970s and 1980s, the exploration, development, and mining for gold and silver, particularly in Nevada, were similar to the gold rush days of the 1850s in California and the Comstock Lode of Nevada of the 1860s except that the prospector and burro were replaced by the Jeep and geologist armed with chemical soil-sampling equipment. The mining companies used for their calculations in 1982-1985 a price of about $400 per ounce, with a production cost of about $200-$250 per ounce to become a viable venture.

It is interesting to speculate that had the price of gold risen in the 1950s to the prices of the 1970s and 1980s the Empire may have been operating today.

The U.S. dollar was devalued to a gold price of $38 per ounce in December 1971 and again in September 1973 to $42.22 for

International Monetary Fund purposes. Since that time the price has fluctuated from below $100 in 1972-73, below $120 in 1976, and again on lows towards $300 in early 1982. The peaks were $200 in 1974, $850 in 1980, and $500 in early 1982.

On December 31, 1974, President Ford ended a 40-year ban on the private ownership of gold by United States citizens.

Gold Theft—Security a Major Problem

SINCE GOLD WAS FIRST DISCOVERED 6000 years ago, the lure and lust for its possession have continued through the ages and it has been considered fair prize for the taking. The story of the Empire must include some comments on Empire Mine thefts. The California early day miners followed the philosophy that gold theft was a miner's right and that taking high-grade gold ore was a just compensation for the hazards and discomforts of working under-ground. Some mines with rich ore pockets had such "highgrading" losses that revenues were seriously handicapped.

Many companies when following a rich gold ore vein would not permit miners to return to the area after blasting in the stope face until the muck-ore pile could be examined by specimen bosses who sorted out visible gold and took it to the above-ground vault.

As an example of the prevalent gold thefts from the Grass Valley mines, the north-south main street of the town was named Mill Street because of the large number of small grinding units (known as coffee grinders) operating after dark grinding high-grade ore to recover the gold. Also contributing to the name of the street was one of the early custom stamp mills in Boston Ravine at the south end of Mill Street. "Highgrading" was not limited to miners underground, but amalgam, precipitates, or other forms of gold were in open season. An example of one form of theft was the practice of suspending a sock full of zinc shavings or dust in a pregnant cyanide leaching tank to collect or precipitate gold, which was then taken home and melted into bullion.

The one positive method for preventing underground highgrading is to use two change rooms for clothes. When a miner comes to work, he leaves his street clothes in one locker room and then passes naked by the security guard into the second room, where he dresses in his underground working clothes. After his shift is finished, he then leaves his mine clothes in the pull up buckets and again walks naked past a security guard into the shower, and dresses in the locker room.

The same method is used in cyanide plants. His lunch pail or bucket is opened and searched by the security guard. Variations and modifications of this method were used at all large mines including the Empire.

In 1934, when gold was declared by the U.S. Government to have a value of $35 an ounce, thievery became rampant. Even with the cooperation of the State Chief Gold Inspector, during the period of 1934 to 1942 the Empire Star never brought to trial miners caught with gold highgrade in their possession for two reasons: one, it was impossible to prove the highgrade orginated from a certain mine, since no chemical analyses could prove this relationship; two, it was impossible to get a jury from a mining camp who believed that a miner should be prosecuted for taking a specimen "rightfully" found underground.

Amalgam was available for theft in the Empire stamp mill-concentrating plant except for the surveillance of trustworthy foremen, some of whom had been employees for 50 years. Also, as a security measure, there was an enclosed elevated walkway from the manager's second story office across the mine yard and into and through the width of the mill, a total distance of about 800 feet. The walkway was equipped with louvered port holes for sight observation throughout the stamp-table concentration area. One could see out but no one could see in. This observation served as a deterrent to theft because the mill hands could be observed day or night.

The one recorded major gold theft occurred in the cyanide plant in the late 1930s. There was a ring comprised of three employees— a shift foreman, shift lead man, and shift operator. It was estimated that they stole perhaps 7000 ounces, then worth about $250,000. The shift foreman, a trusted 15-year employee, altered the weights of concentrate received at the cyanide plant, thereby reducing the tonnage weights, allowing for theft of free gold and amalgam equal to the altered weight shortages. The assay value was not changed. There was a wood stove in the plant office, used in winter for heating. The plant superintendent found small gold beads around the stove which, from the evidence, had been used for melting gold amalgam. This was a tip-off of possible theft and, unbeknown to the culprits, a quiet investigation began of all plant employees. It was soon discovered that the ring of three had set up out-of-town bank accounts and purchased real estate and ranches. They were clever enough to maintain the same mode of living in Grass Valley. It took several weeks of investigative work to trip them with irrefutable evidence. Management decided against prosecution, realizing the futility of a court conviction, and, secondly, admitting that other thefts could possibly occur. The assignment to uncover the method used to disguise the thefts that went undetected for such a period of time was most interesting.

In several years of association with the Mother Lode gold mines and Nevada County mines, no case of prosecution for gold theft is known except recently. In this case the manager-owner of a high-grade vein mine uncovered sufficient evidence of theft by an employee. Over a lengthy period of time, the man was brought to trial and convicted. However, it is doubtful this conviction could have been obtained in the 1930s.

The many tales of gold and silver theft (highgrading) in mines, mills, plants, smelters, and refineries would make an interesting history.

BIBLIOGRAPHY

Bean, E.F. 1867. *Bean's history and directory of Nevada County.* Nevada [City], 424 pp.

Browne, J.R., and Taylor, J.W. 1886. *Reports on the mineral resources of the United States.* U.S. Treasury Dept.

Chandler, J.W. 1941. Mining methods and costs of the Lava Cap Gold Mining Corporation, Nevada City, California, pp. 409-436 *in California Division of Mines and Geology,* v. 37, no 3, 496 pp.

Clark, W.B. 1970. Gold districts of California: *California Division of Mines and Geology Bull., 193,* 186 pp.

Conway, Marion F. 1981. *A History of The North Star Mines,* Grass Valley, Calif., 27 pp.

Crawford, J.J. 1894. Nevada County, pp. 185-203 *in California State Mining Bureau,* v. 12, 541 pp.

Crawford, J.J. 1896. Nevada County, pp. 234-271 *in California State Mining Bureau,* v. 13, 726 pp.

Hague, William, and Pagan, W.D. 1914. The North Star Mine. Grass Valley: *Mining and Scientific Press,* v. 109, pp. 549-552.

Hamilton, Fletcher. 1921. Mining in California during 1920: 17th Annual Report of the State Mineralogist: *California State Mining Bureau,* 562 pp.

Hobson, J.B. 1890. Nevada County, p. 364-398 *in William Irelan, Jr., 10th Annual Report of the State Mineralogist:* California State Mining Bureau, v. 10, 983 pp.

Hobson, J.B., and Wiltsee, E.A. 1893. Nevada County, pp. 263-318 *in 11th Annual Report of the State Mineralogist:* California State Mining Bureau, 612 pp.

Hoover, H.C. 1896. Some notes on "crossings": Mining and Scientific Press, v. 72, pp. 166-167.

Irelan, William, Jr. 1887. Nevada County, pp. 44-52 *in California Journal of Mines and Geology,* v. 6, part 2, 222 pp.

Irelan, William, Jr. 1888. Grass Valley District, pp. 425-435 *in 8th Annual Report of the State Mineralogist,* California State Mining Bureau, 948 pp.

Jenkins, O.P. 1948. The Mother Lode Country (Centennial Edition): *California Division of Mines and Gelology Bull., 141,* 164 pp.

Johnston, W.D., Jr. 1940. The Gold Quartz Veins of Grass Valley, California: *U.S. Geological Survey Prof. Paper 194,* 101 pp.

Koschmann, A.H., and Bergendahl, M.H. 1968. Principal Gold-Producing Districts of the United States: *U.S. Geological Survey Prof. Paper 610,* 283 pp.

Lindgren, Waldemar. 1896a. Nevada City Special Folio, California: *U.S. Geological Survey Folio 29.*

Lindgren, Waldemar. 1896b. The Gold Quartz Veins of Nevada City and Grass Valley Districts, California: *U.S. Geological Survey 17th Annual Report*, pt. 2, pp. 1-266.

Lindgren, Waldemar. 1900. Colfax Folio, California: *U.S. Geological Survey Folio 66*.

Lindgren, Waldemar. 1911. The Tertiary Gravels of the Sierra Nevada of California: *U.S. Geological Survey Prof. Paper 73*, 226 pp.

Logan, C.A. 1930. Nevada County, pp. 90-137 *in California Division of Mines*, v. 26, no. 2.

Logan, C.A. 1941. Mineral Resources of Nevada County: *California Journal of Mines and Geology*, v. 37, no. 3, p. 374-408.

MacBoyle, Errol. 1919. Mines and Mineral Resources of Nevada County: *California State Mining Bureau*, 270 pp.

McQuiston, F.W., Jr., and Keenan, J.C. 1975. Nevada County, California: A Summary Report: Newmont Mining Corp.

Morley, Jim, and Foley, Doris. 1965. Gold Cities, Grass Valley and Nevada City: Howell-North Books, Berkeley, Calif., 96 pp.

Nevius, J.D. 1909. Unpublished report, *The Murchie Gold Mine*, Nevada City.

Raymond, R.W. 1868. Mineral Resources of the States and Territories West of the Rocky Mountains: U.S. Treasury Dept.

Searls, Fred, Jr. 1956. *Empire Star Mines Company, Ltd. Annual Reports, 1929-1956.*

Starr, G.W. 1900. *The Empire Mine, Past and Present:* Mining and Scientific Press, v. 81, p. 120, 152, 184.

Weed, W.H. 1918. *The Mines Handbook:* v. 13, 1896 pp.

Zimmerman, Joseph. 1937. *Mines Register:* v. 19, Mines Publications, Inc., New York.

Glossary of Terms

alluvial. Of or pertaining to alluvium.

alluvium. Sand, mud, or other rock particles deposited by flowing water. Includes sediments deposited both above and below their present water level, such as stream and river bottoms and their banks and flood plains.

amalgamation. 1. The production of an amalgam or alloy of mercury. 2. The process in which gold and silver are extracted from pulverized ores by producing an amalgam, from which the mercury is afterward expelled.

amphibolite. A variety of dark-colored rock. Composed mainly of amphibole and plagioclase. Quartz is either absent or is present in only small amounts. Believed to have been derived from preexisting rocks low in silica by their recrystallization due to deformation.

amphibolite schist. A variety of amphibolite (which see) that has a schistose structure.

andesite. A variety of dark-colored volcanic rock composed essentially of andesine and one or more mineral constituents rich in magnesian rock-forming silicates.

aplite. An igneous rock composed of light-colored minerals and having a fine-grained texture.

arsenic. A native element that occurs as masses of grey hexagonal crystals. This element is usually found combined with other elements to form such minerals as arsenopyrite (FeAsS).

arsenical area. Portions or segments of vein where there is a strong concentration of arsenic.

clastic. A type of rock composed of fragmental material derived from preexisting rocks. A conglomerate such as sandstone.

Cornish buddles. An inclined trough or platform on which crushed ore is concentrated by running water which washes out the lighter, and less valuable, portions.

Cretaceous Period. The third of three major divisions of geologic time within the Mesozoic Era. Extensive layers of chalk were deposited in what is now England during the Cretaceous Period. It is now commonly considered to have lasted approximately 65 million years, from about 135 to 70 million years before present.

cupellation. The treatment on a hearth or cupel (usually formed of bone ash) of an alloy of lead, gold, and silver by means of fusion and an air blast, which oxidizes the lead to litharge and removes it in liquid form or absorbs it in the cupel.

diabase. A rock of basaltic composition, usually occurring in dikes or intrusive sheets.

90

diabase porphyry. A porphyrite whose groundmass is finely crystalline diabase.

dike. Long and relatively thin bodies of igneous rocks which, while in a state of fusion, have entered fissures in older rocks and have there chilled and solidified.

diorite. A granular igneous rock consisting essentially of plagioclase feldspar and hornblende.

dip. The maximum vertical angle at which a stratum or any other planar feature is inclined from an imaginary horizontal plane. This angle is always measured within an imaginary vertical plane that is perpendicular to the strike at that precise location.

electrum. Native gold which contains a small amount of silver.

flotation. Method of mineral separation whereby certain minerals in an aqueous slurry become attached to small bubbles rising through the slurry and are floated off with the froth.

flux. Any substance or mixture that promotes fusion (melting).

gabbro. A variety of igneous rock containing calcic plagioclase (commonly labradorite) and clinopyrozene, with or without orthopyroxene and olivine.

galena. A mineral, lead sulfide (PbS), with metallic luster showing highly perfect cubic cleavage and constituting the principal ore of lead.

gangue. The nonvaluable minerals in an ore.

granodiorite. A term used by the U.S. Geological Survey for intermediate rocks between granites and quartz-diorites. It is a contraction of granite-diorite. Cretaceous Granodiorite—granodiorite formed in the Cretaceous Period preceding the Tertiary, 70 million to 135 million years before present.

hydrometallurgical. The practice of extracting metals from ores by leaching with solutions such as cyanides, acids, brines, etc. Wet extraction methods in general.

igneous. Molten or partly molten minerals as well as the rocks formed by solidification of such a magma. The word is derived from the Latin meaning "fire" in reference to heat, an essential ingredient in the development of these rocks.

leach (leaching). To remove soluble substances from something by percolating water or other dissolving solutions through the mass.

lode gold. A fissure in the country-rock containing a gold-bearing ore. A vein and a lode are, in common usage, essentially the same thing. The filling of a fissure or fault in a rock, particularly if deposited by aqueous solutions. When metalliferous it is called by miners a *lode;* when filled with eruption material, a *dike.*

magma. Molten material beneath the solid crust of the Earth, from which igneous rock is formed.

magmatic. Rocks formed by cooling of a magma.

moil. 1. A short length of steel rod tapered to a point. Used for cutting hitches (pockets for holding the end of a timber). 2. A long gad (steel wedge) used for accurate cutting in a mine.

orogeny. The process of forming mountains, particularly by folding and thrusting.

parting of gold and silver. Separation of gold from silver in a gold-silver alloy by means of nitric or sulfuric acid.

Pelton Wheel. After Lester Allen Pelton—inventor. A water wheel consisting of a row of double cup-shaped buckets around the rim of a wheel and actuated by one or more jets of water playing into the cups at high velocity.

placer. A place where gold is obtained by washing; an alluvial or glacial deposit, as of sand or gravel containing particles of gold or other valuable minerals.

porphyry. A variety of fine-grained igneous rock containing conspicuous phenocrysts (crystals) of alkali feldspar.

pyrite. A hard, heavy, shiny, yellow mineral, iron sulfide (FeS_2), generally in cubic crystals. Also called iron pyrite, fool's gold. The plural, pyrites, is applied to any number of metallic-looking sulfides such as copper pyrites, tin pyrites, etc.

quartz. 1. A mineral having the chemical formula SiO_2. Its crystals are commonly hexagonal prisms that are striated crosswise. It breaks with a conchoidal fracture and has a high luster. It is the seventh of ten minerals in Moh's scale of relative hardness. 2. (Along the Pacific Coast) Any hard gold or silver ore, as distinguished from gravel or earth. Hence, quartz mining as distinguished from hydraulic mining.

Quaternary Period. The younger of two geologic periods of time in the Cenozoic Era. Geologists presently regard it as the period of time beginning about 2 million years ago and extending to the present time. This period includes the Ice Age.

rhyolite. A light-colored variety of volcanic rock containing one or more silicate minerals, usually quartz. It resembles granite in its mineralogical and chemical properties and is considered to have been the extrusive equivalent of a magma composed of molten granitic rocks.

rhyolite andesite tuff. See *rhyolite, andesite,* and *tuff.* A variety of tuff composed of rhyolite and andesite.

riffles. From the Danish *rifle,* a groove or channel. In mining, the lining of the bottom of a sluice, made of blocks or slats of wood or stones arranged in such a manner that chinks are left between them. The whole arrangement is usually called *the riffles.* The slots nailed across the bottom of a cradle are called *riffle bars* or simply *riffles.*

schistose. A medium- or coarse-grained metamorphic rock in which its micaceous minerals possess a sub-parallel orientation. These rocks have a foliated structure which allows many of them to be split into thin irregular plates.

serpentine. A class of ferromagnesian silicate minerals formed by alteration of previously existing olivine, pryoxene, or other minerals low in silica. Includes at least two distinct minerals, antigorite and chrysotile, the latter including asbestos.

shallow vein. As used here, meaning veins just below the surface.

slate. A fine-grained metamorphic rock that can be split into thin parallel sheets. This rock is commonly derived from preexisting shale.

slime fraction. That portion or fraction of a product of wet-crushing containing valuable ore in particles so fine as to be carried in suspension by water. In metallurgy that part of an ore reduced to a very fine powder and held in suspension in water so as to form a kind of thin ore-mud.

sluice boxes. A wooded trough in which alluvial beds are washed for the recovery of gold or tinstone.

stope or stoping. Stope is an excavation from which the ore has been extracted, either above or below a level (usually above) in a series of steps. To excavate a vein by driving horizontally upon it a series of workings, one immediately over the other. Each horizontal working is called a stope (probably a corruption of step) because when a number of them are in progress, each working face a little in advance of the next below, the whole face under attack assumes the shape of a flight of stairs.

strakes (Cornish). 1. A trough for washing broken ore, gravel, or sand. A launder. 2. The place where ore is assorted on the floor of a mine, a dressing floor.

strike. The course or bearing of an inclined bed or stratum, fracture, or other structural feature at a specific place in space. In practice, geologists describe this bearing by giving the direction of an imaginary horizontal line within the plane of the inclined bed or stratum. This direction is usually given in degrees east or west of true north. Example: N 35° W means that the surface being measured contains an imaginary horizontal line that trends approximately 35° west of true north.

stulls. A round timber used to support the sides or back of a mine. A series of props wedged between the walls of a stope to hold platforms for miners.

telluride. A mineral containing tellurium.

Tertiary Hills. Hills formed during the Tertiary Period of the Cenozoic Era, approximately 2 million to 70 million years ago. The period during which the Pacific (Cascadian) and Rocky Mountain Ranges began to form.

tuff. A rock formed of compacted volcanic fragments which are generally smaller than 4 mm. in diameter. The deposit is commonly formed by an ash fall, that is, a rain of airborne volcanic ash falling from an eruption cloud.

vein. Crack in bedrock largely filled with mineral matter deposited by mineralized ground water.

winze. A vertical or inclined opening or excavation connecting two levels in a mine, differing from a *raise* only in construction. A winze is sunk underhand and a raise is put up overhand. When the connection is completed, and one is standing at the top, the opening is referred to as a winze, and when at the bottom, as a *raise* or *rise.*

ABOUT THE AUTHOR

Frank Woods McQuiston, Jr., was born in Pueblo, Colorado, and spent his youth in mining camps of Colorado, Utah, and Arizona. He received a Bachelor of Science degree from the College of Mines, University of California, Berkeley, in 1931.

In 1951 he was awarded the French "Order of Ouissam Alaourite Cherifien" for outstanding metallurgical services to the Moroccan Government. He was presented with the Robert H. Richard award in 1968 by the American Institute of Mining Engineers; made a Distinguished Member of the Society of Mining Engineers Class of 1975, and an Honorary Member of AIME 3 years later.

He was a prolific contributor to the literature of mineral processing and metallurgy including the AIME volumes *50th Anniversary of Froth Flotation, Surface Mining, Mining Engineers Handbook,* and the *Mineral Processing Handbook.* He also was the coauthor of two monographs entitled *Gold and Silver Cyanidation Plant Practice* and another, *Primary Crushing Plant Design.*

Mr. McQuiston retired from Newmont Mining Corporation in 1969, but remained as a consultant until 1982.